Remediation of Soils and Aquifers

REMEDIATION OF SOILS AND AQUIFERS

LUIS G. TORRES
AND
ERICK R. BANDALA
EDITORS

Nova Science Publishers, Inc.
New York

For permission to use material from this book please contact us:
Telephone 631-231-7269; Fax 631-231-8175
Web Site: http://www.novapublishers.com

LIBRARY OF CONGRESS CATALOGING-IN-PUBLICATION DATA

Remediation of soils and aquifers / Luis G. Torres and Erick R. Bandala, editors.
p. cm.
Includes index.
ISBN 978-1-60741-372-1 (hardcover)
1. Soil remediation. 2. Soil pollution. 3. Water--Pollution. 4. Water--Purification. 5. Aquifer storage recovery. 6. Wellhead protection. I. Geng T., Luis. II. Bandala, Erick R.
TD878.R4587 2009
628.5'5--dc22

2009018005

Published by Nova Science Publishers, Inc. ✦ New York

CONTENTS

FOREWORD

The cost of cleaning up water and soil leads to the same undeniable conclusion: Existing models of industrial development are simply not sustainable from an ecological view point and they cost us a lot of money in purely economic terms. Mexico is no exemption to this general rule and soil and groundwater contamination are neither here, nor in other Latin American countries a new or emerging issue. As an oil exporting nation, Mexico, historically has paid major attention to hydrocarbon contamination and its implications for the environment, especially when its impact was related to drinking water safety and public health criteria. It is only more recently, that a more generic and holistic approach has emerged, which has included other industrial sectors and their environmental liabilities. This new approach is directed towards the government´s and private industry´s extended obligation and responsibility in mega-site remediation, brown-field re-vitalization and the re-urbanization of degraded land.

A great step forward for Mexico has been the ratification of the Waste Management Act (Ley General para la Prevención y Gestión Integral de los Residuos, LGPGIR) in 2004 and more specifically the outlines given in its Chapter V, which regulates the management of contaminated sites. Subsequently, the ratification of the ordinance to LGPGIR put in place in 2006 and a couple of technical norms and regulations, like NOM-147 limiting heavy metals content in soil and NOM -138 regulating hydrocarbon presence in soil, to name the most prominent, have set the pace for an orderly and legally sound approach to brown-field remediation.

Other public policy instruments include the Federal Environmental Ministry´s National Remediation Plan for Contaminated Sites presented in December 2008. Additionally, SEMARNAT´s internal information system for contaminated sites –SISCO, in place since 2007, enables the government to enforce remediation action based on priority listings for severely contaminated sites in various states of the Mexican Federation.

The academic sector is also involved and this book is proof of it. The redevelopment of brown-field sites requires a full assessment of all environmental factors, such as the potential soil and water contamination and the health risk emanating from them. Remediation strategies and techniques can be designed once a full understanding of a site is known. Various chapters of this book are dedicated to these issues.

This book arose from the scientific interest and the environmental preoccupation of researchers and governmental representatives alike. They provide a concise overview of concepts and remediation/sanitation technologies available today. At the same time they bear witness to the wealth of solutions available in order to tackle the inherent and very serious problem of environmental degradation in Mexico. Thus, this book will serve as a useful source of technical information as well as guidance document for state and municipal administration and regulation staffers, researchers, environmental consultants and any interested person with environmental responsibility.

Dr. Wini Schmidt-German Agency for Technical Cooperation (GTZ)

In: Remediation of Soils and Aquifers
Editors: Luis G. Torres and Erick R. Bandala
ISBN 978-1-60741-372-1

Chapter 1

NATURAL ATTENUATION AND PHYTOREMEDIATION OF PETROLEUM HYDROCARBON IMPACTED SOIL IN TROPICAL WETLAND ENVIRONMENTS

***Randy H. Adams*[1]*, Ofelia Castillo-Acosta*[1]*, Erika Escalante-Espinosa*[1]*, and Joel Zavala-Cruz*[2]**

[1]Universidad Juárez Autónoma de Tabasco
[2]Colegio de Postgraduados, Campus Tabasco

ABSTRACT

The natural processes that mitigate or eliminate hydrocarbon caused impacts to floodable soils in a tropical environment are qualitatively different from those in upland soils. These differences are important to understand when trying to employ natural low cost attenuation or phytoremediation methods to remediate these kinds of sites. These differences are due mainly to the anaerobic nature of floodable soils and the relatively high production rates of plant biomass in these kinds of ecosystems. These considerations are especially important since a large part of the petroleum production in some tropical areas is in marshes, floodable forests, and floodable pasture lands (for example in SE Mexico, the Niger Delta in Nigeria, and the Orinoco Delta and part of the Llanos region of Venezuela), and these same arguments could also apply to the very high petroleum production area in the semitropical climate of the Mississippi Delta in Louisiana (U.S.A.). In this paper we present the natural mitigating phenomena which occur in these environments, phytoremediation potential, and restoration schemes, especially by biological stabilisation and confinement processes.

Keywords: marsh, mangrove, oil, remediation, toxicity, stabilisation, silviculture.

INTRODUCTION

In the last 10 – 15 years in various parts of the world, people have become aware that environmental remediation and restoration programmes are often over ambitious, in that they try to remediate contaminated sites in a very short time, but at a cost which can not be justified in terms of reduction in human health risks, betterment of agricultural or livestock production, or in terms of recovered ecological value (Carman *et al.*, 1997; Hinchmann *et al.*, 1997; Woodward and Day, 1996). At many sites, there are natural processes which can play a role in environmental impact mitigation and restoration, over a much longer time, but at much lower, justifiable costs. For upland soils, the natural processes which mitigate the risks associated with hydrocarbon contamination have been well studied (Zavala *et al.*, 2002; Schnoor *et al.*, 1995), but in marshy areas there is relatively little understanding (Mendelssohn *et al.*, 1996). The objective of this paper is to explain how these processes work in marshy environments and how they can be used for natural attenuation and phytoremediation.

Comparison Between Intensive and Moderate Technologies

Many people believe that the future of remediation, especially for developing regions, will rely heavily on treatment methods which are low-cost and of extended durations. In almost all remediation or restoration programmes, longer treatment times cause the site to be unusable or less usable for economically productive activities for longer periods, resulting in a greater loss of economic opportunity and gains. Therefore, the higher cost of faster, more intensive technologies is justified when the economic utility of a property is high, such as is the case for many urban and industrial settings. In these kinds of settings (urban), the high cost of intensive technologies with shorter treatment times is offset by a reduction in the loss of economic opportunity that would occur using a lower cost, but longer duration treatment technology. The result is an overall reduction in costs (including both the remediation itself and the lost economic opportunity and gains).

However, for sites in rural environments (as are most petroleum production sites) the opposite is true. In these areas the economic utility (amount that can be earned using the land) is less, and therefore low cost, longer duration remediation methods are preferable. The longer treatment time typically causes the site to be unusable for economically productive activities for much longer, but this loss in economic opportunity is much less than the higher cost of more intensive, shorter duration technologies. The overall costs, including the remediation and lost economic gains, are reduced (see Table 1).

The more moderate treatment technologies in terms of cost and logistics are phytoremediation and natural attenuation. In this paper we present some of the natural processes which work to mitigate hydrocarbon contamination in marshy areas which can be incorporated into restoration programmes using these technologies.

Table 1. Applications of Intensive and Moderate Remediation Technologies.

Theoretical situation	Economic opportunity	Appropriate technologies	Observations
Contaminated soil underneath a gasoline station in a commercial plaza in a large city, such as Mexico City	Every day that the station is closed a large potential for economic gain is forgone	Intensive: Excavation and con-finement in an engineered landfill	Site closure for a minimum amount of time, transport and landfill costs, backfill costs
		In situ bioremediation	It may not be necessary to close the gasoline station but certain incon-veniences must be tolerated for a longer period of time, the cost is less but there may be geological limitations
		Excavation and thermal desorption	Site closure for a relatively short period, cost is less than the first option
Hydrocarbon spill in a oil field in a rural area, such as in southern Veracruz state (Mexico)	The economic opportunity is much less than that in an urban or industrial area	Medium: Bioremediation (land farm)	It is the most expensive in this category, but is faster
		Composting	The cost is low if the handling can be kept to a minimum, it is effective
Old oil spill in a cattail marsh in western Tabasco state (Mexico)	The economic utility of the site is practically nil.	Moderate: Phytoremediation with cattail (*Typha lati-folia)* and/or a species of mangrove	Low cost, treatment time very long, but it is effective
		Natural attenuation	Minimum treatment costs (characteri-sation, modelling, monitoring), it is effective but has a very long treatment time, there may be some environ-mental limitations

Natural Processes Involved in Mitigation and Restoration in Marshy Environments

There are various vegetative mechanisms or processes that are involved in the restoration or mitigation of soil and water impacted by petroleum hydrocarbons in marshy areas. Some of the most common are described below (Adams and Castillo, 2000):

Adsorption/Absorption

Plant roots are capable of adsorbing and/or absorbing various kinds of organic compounds including many chemical species of hydrocarbons. This reduces their bioavailability and toxicity in the soil, and in some cases permits their subsequent transpiration or biochemical stabilisation, as well as decomposition of the hydrocarbons by microbial and vegetative enzymes.

Various researchers (Burken and Schnoor, 1997; Al-Assi, 1993; Briggs, et al., 1982) have shown that the hydrocarbons that have a greater tendency to be absorbed by roots are the less polar kind, especially polyaromatic hydrocarbons, and that in general, these are not transported to plant surface tissues. Furthermore, the absorption in roots is proportional to the octanol-water partition coefficient (K_{ow}) for that compound. When the K_{ow} is higher (such as in heavy oil range hydrocarbons), the absorption to the root is greater. Less polar compounds have a greater tendency to be absorbed to the root, but if the compound is very lipophilic, it will not pass from the root into the plant xylem (sap), and is not transported from the roots into plant surface tissues, such as shoots and leaves. In practical terms, semivolatile and non volatile hydrocarbons tend to be tightly absorbed to roots and are not transported to plant aerial structures.

This has important consequences for phytoremediation projects, as well as for natural attenuation in marshy areas. It is very improbable that the shoots and leaves of pasture or other herbaceous plants will be affected by residual hydrocarbons in soil, and thus the risks to wildlife that is sustained by this vegetation is reduced, as well as that to cattle which may forage in nearby areas.

Also, the absorption of hydrocarbons to the soil organic material (SOM) is very important in marshy areas due to the large quantity of it in these kinds of soils (up to 75 – 86 % organic material; Moreno, 1999). Likewise, the K_{ow} values for semi- and non-volatile hydrocarbons are very high, from values of approximately 10^6 for the lighter diesel range hydrocarbons (semi-volatile) up to values higher than 10^9 for heavier fractions (Wyoming Dept. Environ. Qual., 2003; Gustafson *et al.* 1997). This means that the partitioning of these compounds between the water and SOM in a marshy area is on the order of less than one in a million. This level frequently is not toxic for these kinds of compounds (Olán, 2000; Bolaynas, 2003; Gustafson *et al.* 1997; Edwards *et al.* 1997; Vorhees *et al.* 1997), and it is much lower than that permitted by many regulatory agencies (Adams, 2000). This mechanism (absorption) effectively immobilizes the majority of hydrocarbons from a spill.

Transpiration

Some of the lighter, gasoline range hydrocarbons, as well as many solvents can be absorbed by plant roots and from there be transpired through the leaves to the atmosphere. This in effect reduces the concentration of these kinds of hydrocarbons in a marshy area. Normally, this process does not occur so fast as to cause significant pollution problems in the atmosphere. The potential of a hydrocarbon to be transpired appears to be proportional to its polarity; those compounds with K_{ow} between 10^2 and 10^3 (such as benzene, toluene or m-xylene) tend to be those with greater potential to be transpired (Burken and Schnoor, 1997).

For marshy areas, this process is only important for very recent hydrocarbon spills with very light fractions. None-the-less, between this mechanism (transpiration), volatilization, and the stimulation of microorganisms in the rhizosphere, the lightest and most toxic hydrocarbons from a recent petroleum spill could be eliminated from a marshy area relatively quickly.

Rhizosphere Biodegradation and Mineralization

The root zone (rhizosphere) is the part of the soil around the root which is modified chemically and physically by plant roots, and is rich in organic compounds called root exudates (Singer *et al.* 2003). These compounds normally stimulate the nearby microorganisms, and often increase the microbial activity between 20 and 1000 times (Atlas and Bartha, 1987). Furthermore, plant roots physically open up the soil and increase moisture infiltration, bettering internal drainage and gas exchange (aeration, Erickson et. al., 1993). Perennial grasses are especially useful in this capacity (Grey, 1994; Aprill and Sims, 1990). The sum of these processes stimulates soil microorganisms, frequently increasing the rate of hydrocarbon decomposition.

In marshy areas these processes probably only occur in the first few centimetres of the soil surface. Due to flooding, soil conditions are generally anaerobic, and the decomposition rate of organic matter (of vegetable origin or from petroleum) is much reduced. None-the less, this mechanism (stimulation of petroleum degrading microorganisms) is important in the detoxification of the most toxic, lower molecular weight, hydrocarbons common in fresh spills (of less than a year old) on the marsh surface, and subsequent ecosystem recovery.

Decomposition due to Root Enzymes

Roots produce enzymes which partially decompose certain hydrocarbons (Drake, 1997a). It appears that some such enzymes, especially in grasses, are complementary to microbial enzymes, and accelerate the decomposition of some hydrocarbons, especially the polyaromatic fraction (polyaromatic hydrocarbons, or PAHs). This is especially important due to the difficulty that many microorganisms have in degrading PAHs and due to the carcinogenic character of some of these compounds. These enzymes include laccases, dehalogenases, nitroreductases, nitrilases and peroxidases (Salt *et al.* 1998).

This process helps in marshy areas to oxidize and partially decompose the heavier molecular weight hydrocarbons found in the first few centimetres of surface soil. At lower depths the anoxic conditions in the marsh do no permit high degradation rates, but on the surface these compounds become more decomposed and more easily sequestered in humification processes (see following section).

Biological Stabilisation (Humification)

A process occurs in the rhizosphere in which hydrocarbons, as well as plant derived organic matter, are partially decomposed by plant and microbial enzymes, and latter undergo a spontaneous chemical condensation reaction which links them into an irregular form (humus, Schwab, 1997). In this form, the partially decomposed and oxidized hydrocarbons are incorporated on the molecular level into the humus. Hydrocarbons sequestered in this way present very low bioavailability, greatly reducing their toxicity and reactivity (they are practically inert). This biological stabilisation is shown in Figure 1. This process is very similar to microbial humification except that root enzymes also play a role in the preliminary transformations of the hydrocarbons, especially in the case of polyaromatic hydrocarbons.

Since this mechanism is aerobic, it is only important in the first few centimetres of marshy soil, but it does help stabilise and detoxify surface crusts.

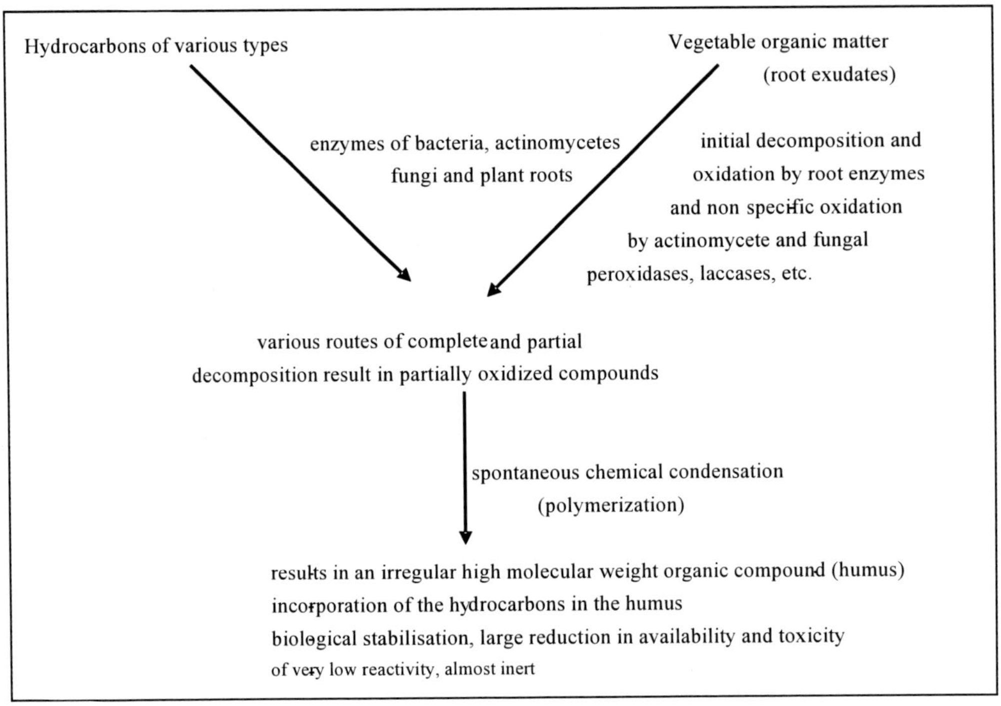

Figure 1. Biological Stabilisation of Hydrocarbons in Humus.

Natural Biological Confinement

Different kinds of natural processes are found in diverse ecosystems which are important in recovery or mitigation of a contaminated site. The conditions found in marshy environments are especially important since many land-based petroleum production areas in the world are situated in deltas and floodable environments (such as southern Louisiana, SE Mexico, the Niger River Delta, the Orinoco River Delta, etc.). Particularly, for example, in the petroleum producing state of Tabasco in SE Mexico, approximately half of the territory is floodable, and nearly 90 % of the oil installations are in marshy areas. It is in these kinds of ecosystems that a natural recovery process of "biological confinement" has been observed, which appears to be the most important process in floodable areas.

In many marshes, mangroves and other floodable forests, the primary productivity (vegetative biomass production) is very high (Atlas and Bartha, 1987). This is because: 1) these areas receive nutrient rich runoff from nearby (often alluvial) soils, 2) moisture is almost never limiting, and 3) at least in tropical and subtropical environments, the solar energy available for photosynthesis is abundant.

Added to this phenomenon (high biomass production), naturally low decomposition rates of organic matter is also found in these environments, due to the anoxic conditions common in marshes. The combination of high primary productivity and low biomass decomposition results in the accumulation of organic matter in these ecosystems, often at relatively rapid rates (especially in the tropics).

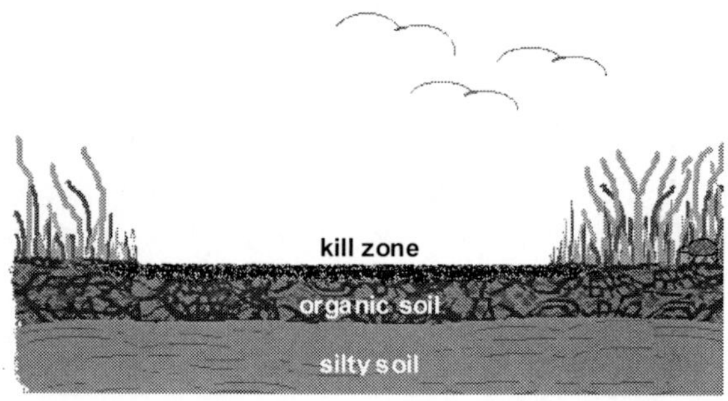

Recent Oil Spill in a Marsh. In a new spill, the lighter fractions of crude petroleum are very toxic and kill the vegetation (and fauna) in the spill area and nearby. Native fauna which survive migrate to adjacent areas.

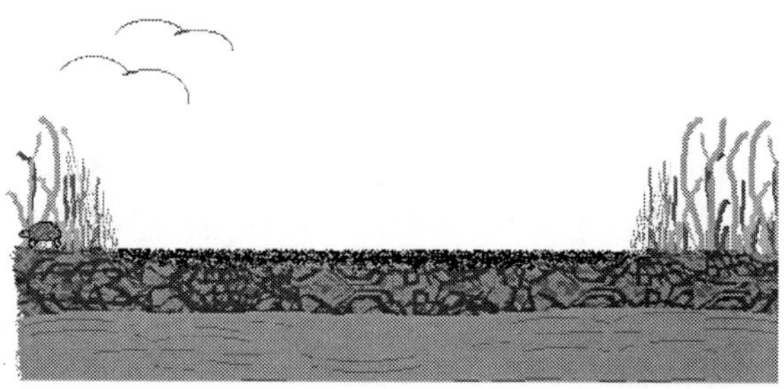

Interment of the Spill. Over the years a considerable quantity of fresh leaf litter is converted into organic soil, effectively burying the spill and forming a new layer of uncontaminated soil.

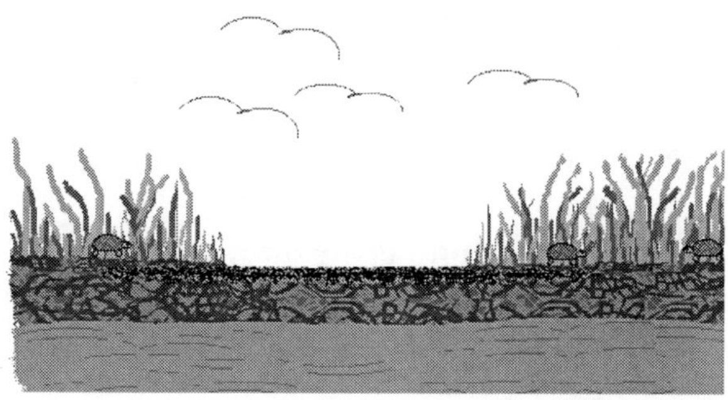

First Colonisers. On the edges of the spill some of the more resistant plants start to grow, forming new leaf litter, very little of which is decomposed due to the anoxic conditions in the marsh. The result is the formation of new organic soil on top of the spill. This same process results in the accumulation of organic matter in the unaffected marsh area, increasing the thickness of organic soil in the overall area. Some fauna start to return to the area.

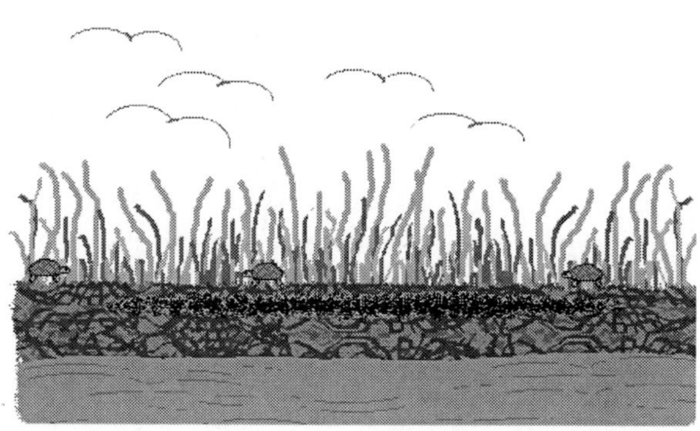

Recovery. Over many years, it is possible to completely cover the spill with a thick layer (approximately 30 cm) of organic soil and vegetation, resulting in the biological confinement of the hydrocarbon impregnated soil layer, and significantly reducing the exposure to the native flora and fauna. With more time the ecosystem becomes diversified in terms of floristic composition, and wildlife.

Figure 2. Biological Confinement in Marshes by Succesional Colonization and Peat Production at Edges of Oil Spill (Cinco Presidents Petroleum Production Unit, Tabasco State, Mexico).

In the long term, this organic matter accumulation results in the burying of oil contaminated soil with fresh, uncontaminated material. The oily soil remains always further from the more recent surface horizons and continually less contact with the native flora and fauna (or introduced species). In this respect, this process is a kind of “biological confinement”, which reduces the possibilities of interaction between sources of toxic compounds and potential receptors in the ecosystem.

Three ways in which this natural biological confinement of oily soil or sediment occurs have been observed. Typically, near the edges of a spill area, vegetation which is more tolerant of heat, drought and low fertility common in a spill (such as grasses and cyperaceas, Gallegos-Martínez *et al.* 2000) become established (Adams *et al.* 1999). This vegetation grows in the less contaminated soil at the edges of the stained, oil impregnated area, but part of the leaf litter produced falls on top of the oily crust, forming a thin layer of organic soil. This mitigates the extreme conditions of the spill area, reducing surface temperature, and slightly improving moisture retention and nutrient supply, which subsequently permit less tolerant species to colonize this area. Consequently, the “kill zone” (oily stained area without vegetation) is reduced, and the edge of the spill moves closer to the centre of the spill. On the new edges the more tolerant species move in, but behind them slightly less tolerant species may start to colonize the recovered area, and with greater vigour. The result is a succession in vegetative colonisation, leaving the oily stained area continually reduced until it is completely covered by fresh organic soil and marshy vegetation (see Figure 2).

The second process of natural confinement which has been observed is by means of vegetative reproduction of cattails (such as *Typha latifolia*). This species is able to extend its roots laterally underneath oily surface crust, and at a distance of approximately 40 cm, it starts to form a root node to establish a new plant (see figure 3).

The new plant starts its growth by perforating the oily crust with the point of the new leaves in development. In this way the plant obtains CO_2, O_2 and energy from the surface, and its moisture and nutrients from below the oily crust, which at this stage is not so limiting to its development. Cattails have been observed to extend themselves from the edge of an oily crust towards the interior of an oil spill at a rate of approximately 10 m a year by this mechanism, invading oil spills. Besides simply extending themselves, they fill in the space between the plants with new individuals, and the leaf litter produced on top of the oily crust forms a new organic soil which mitigates the extreme conditions on the oil surface. This allows the establishment of other herbaceous species on top of the oily crust, and even permits the germination and survival of mangrove seed plantules (*Laguncularia racemosa*) in areas in which both species (cattail and mangrove) are present (see Figure 3).

Lastly, in some old spills, a layer of organic soil can be formed on top of an oily crust due to algal growth. It has been observed that during flood and high water, algae grow on top of these crusts, forming a mat a few millimetres thick. During the dry season, this mat dries completely, and the algae die, leaving behind organic matter which becomes a new organic soil. After a few decades, sufficient soil is produced to allow the establishment of a few of the more resistant species (such as grasses and cyperaceas) which accelerate the process of biological confinement (Adams *et al.* 1999).

For a comparison of natural processes involved in site restoration in marshy areas and upland areas see Table 2.

Figure 3. Biological Confinement by Vegetative Reproduction of Cattails (Texistepec, Veracruz State, Mexico).

Table 2. Comparison of Natural Attenuation Mechanisms in Marshy Areas and Uplands.

Natural Attenuation Mechanism	Upland Areas (with good drainage)	Marshy Areas
Adsorption/absorption	Important for limiting the migration of contaminants toward the water table or to immobilise contaminants in the saturated zone	Very important for immobilising contaminants and reducing horizontal migration in a marsh; this process is greatly increased due to the high concentration of organic material in wetland soils
Transpiration	Mechanism which reduces the concentration of low molecular hidrocarbons in the saturated zone	Of limited importance; however these mechanisms help to eliminate the lightest and most toxic hydrocarbons from marshy areas in a very recent spill
Stimulation of hydrocarbon biodegrading micro-organisms	Very important for the decomposition of superficial contaminants and for reducing the probability that they migrate to the water table	

Table 2. Continued

Natural Attenuation Mechanism	Upland Areas (with good drainage)	Marshy Areas
Descomposition due to root enzymes	Helps in the early decompostion of polyaromatic hydrocarbons (some of which are cancerous) in surface soils with grasses	These aerobic processes are of lesser importance in wetland areas, but they can be significant in the stabilisation and detoxification of oily surface crust in larger spills
Biological stabilisation (humification)	This is the principal mechanism of detoxification of many compounds which are dificult to biodegrade	
Natural biological Confinement	Does not occur, or only very slowly, in areas with good drainage, mineralization and stabilisation (humification) being much more important	This is probably the main mechanism of recovery in marshy areas affected by larger spills; it occurs relatively fast due to the high primary productivity (biomass) and low biodegradation (of leaf litter) en these ecosystems. It is much faster than biodegradation or stabilisation in anaerobic environments.

Phytoremediation in Marshy Areas

Several projects and investigations on the impacts of petroleum in marshy environments have been realized in recent years. Below a brief description of some of the more relevant studies is presented (Adams and Castillo, 1998).

Oil Spill-Caused Impacts to Mangroves

Among the most important factors are the tolerance of mangrove species to anoxic conditions in the soil, and their sensitivity (or lack thereof) to hydrocarbons (Clark, 1996; Glooshenko, 1996; Van Fleet, 1994). It appears that mangroves may be sensitive to recent spills, probably due to the lighter molecular weight hydrocarbons, especially the mono-aromatics and short chain alkanes. In other occasions mangroves have been observed to be resistant to high concentrations of hydrocarbons. Apparently, this occurs with the less toxic fraction of hydrocarbons, usually after a period of weathering, and when the concentration of low molecular weight hydrocarbons have been reduced (due to volatilization and/or biodegradation). White and black mangroves (Laguncularia racemosa and Avicenna germinans, respectively) are relatively resistant to the toxicity presented by hydrocarbons while red mangrove (Rhizosphera mangle) is particularly sensitive (Duke et al. 1997).

One important mechanism whereby mangroves are destroyed by hydrocarbons appears to be asphyxia of the pneumatophors (aerial roots). This occurs in catastrophic spills when the oil covers these parts of the plant, and gas exchange (CO_2 and O_2) is hindered. After a mangrove kill off, new individuals (especially white and black mangroves) are able to recolonise an affected area typically within two years, if the quantity of oil spilled is not too great (Duke *et al.* 1997).

Tolerance of Neotropical Plant Species to Hydrocarbon Contamination

The success of the phytoremediation process requires the colonization of plants on the contaminated soil (White *et al.* 2000). Those plants native to a contaminated site offer key advantages due to the fact that they are adapted to site environmental conditions. In many herbaceous plants, the selection pressure of the contaminants has resulted in the natural evolution of plant genotypes which are tolerant to a large quantity of contaminants in air, soil, and water. The presence of tolerant herbaceous plant species is favoured by natural selection in contaminated environments due to their ability to survive and compete, excluding the non-tolerant plants (Dickinson *et al.* 1991).

On the other hand, there are few studies on native plants that have been carried out in the tropics. Gallegos-Martínez *et al.* (2000) determined those vegetable species present in a tropical marsh contaminated with hydrocarbons located in southeast Mexico. At this site, six dominant species were found (*Thalia geniculata, Phragmites australis, Typha dominguensis, Cyperus papirus, Cyperus laxus* and *Echinocloa colonum*) to be related to the presence of high concentrations of hydrocarbons int the soil. At a concentration of 10 000 to 100 000 mg TPH Kg^{-1} of dry soil associations of Cyperaceae – Gramineae – *Typha dominguensis* were found, while in areas contaminated at a concentration of 100 000 to 434 000 mg TPH Kg^{-1} of dry soil, only the Cyperaceae – Gramineae association was observed. Subsequently, greenhouse studies were run with *Cyperus laxus*, to evaluate the effect of adding hydrocarbon degrading microorganisms with respect to the phytoremediation of hydrocarbons in model system. The results showed that the plant is able to eliminate the hydrocarbons from the medium and that the addition of an inoculum improved the rate of phytoremediation, achieving a removal of 87 % the hydrocarbons after 180 days. Furthermore, the phenological characteristics of the plant (overall plant and root growth), were improved in the inoculated plants, and the early onset of flowering was also observed (Escalante-Espinosa *et al.* 2005).

Recently, Merkl *et al.* (2004) carried out studies on the selection of eight potentially phytoremediating species from eastern Venezuela, based on the effect the crude petroleum produced on growth. The species tested where *Brachiara brizantha, Panicum maximum, Calopogonium muconoides, Centrosema brasilianum, Desmodium glabrum, Mimosa orthocarpa, Mimosa camporum* and *Stylosanthes capitata*. In this study the authors concluded that hydrocarbons have a greater effect on plant growth than on germination, selecting *B. brizantha, C. Brasilianum* and *C. muconoides* as the species with greater potential to be used in phytoremediation projects.

Phytoremediation

Phytoremediation is a relatively new technology in development. The species most studied to date for remediation of hydrocarbons are grasses and some "weeds", which improve the conditions of the surface soil for biodegradation of hydrocarbons by bacteria (Grey, 1994; Drake, 1997; Jackson, 1997; Rodríguez, 1996; Zavala *et al.*, 2002). Deep rooted trees have also been investigated which clean up aquifers by transpiration of contaminants founds in relatively low concentrations, and which control the hydrology of the water table (Anderson and Coats, 1994; Carman, 1997).

In the tropics there have been few phytoremediation projects, but some have been realised in Costa Rica through the nationalized oil refinery company, Refinación Costarricense de Petróleo, Recope. This company has mitigated problems with drilling cuttings (from exploration wells) using various plants, including species of eucalyptus. The high transpiration rate of these trees reduces the groundwater migration away from the affected area, thus reducing the risk of contamination of nearby areas. This, combined with the tolerance of these trees to oily contaminants, allows them to be used for mitigation (Chávez, 1996).

It is also important to mention that there have been some reforestation projects of hydrocarbon affected mangroves, however, the results in terms of detoxification or hydrocarbon concentration reduction were not evaluated. Two of these projects were carried out in the Island of St. Croix, in the Virgin Islands (USA) and also near Colon on the Caribbean coast of Panama. In these two projects the principal objective was to re-establish the mangrove ecosystem, and apparently no evaluation of the potential to use mangroves for phytoremediation (in terms of toxicity reduction and confinement in sediments) was made (Duke *et al.*, 1997; Lewis, 1990). Another project worth mentioning included the sowing of red mangrove (*Rhizosphera mangle*) plantules behind the gas refining plant in La Venta (Tabasco State, Mexico), which was not successful, probably due to the sensitivity of this species to hydrocarbon toxicity (Pérez, 1998; Duke *et al.*, 1997).

Biological Confinement in Marshy Areas

Low-lying hydrocarbon affected areas can be recovered using vegetation typical of marshes or floodable forests (swamps, such as mangroves, for example), which in the long term, inter the contaminated material with new organic soil, and thereby drastically reduce the biological availability of the potentially damaging compounds. Frequently, especially in old spills, the remaining hydrocarbons are very weathered and low in toxicity (see Table 3). In these cases the limiting factor for ecosystem recovery is not toxicity but the physical obstruction of the oily crust between the vegetation and subsoil. Apparently, this does not permit access to the soil moisture and nutrients underneath the oily crust (Adams *et al.* 1999).

To rectify this situation, the nationalized oil company of Mexico, Petróleos Mexicanos, or Pemex, has realised some re-vegetation projects in marshy areas in recent decades (Pérez, 1998). In one case marshy vegetation (cattails, *Typha latifolia*) was planted after the oily crust was broken up, behind the La Venta Gas Processing Plant, Tabasco, Mexico. After a few years, a vigorous plantation was established, such that the oil stained soil below was not even visible. During the regional drought in 1998 this area caught fire and the vegetation was completely burned, leaving only ashes and an oily crust. Five years latter, the vegetation had grown back exuberantly, dominated by cattails, which reached 2 ½ metres high and completely covering the oily crust. Recent observations have shown an accumulation of approximately eight to ten centimetres of fresh organic material deposited on the surface of the oily crust since the fire, indicating an accumulation rate of roughly 1 ½ to 2 centimetres per year. Most agronomist consider that the 0 – 30 cm root zone to be most important for soil fertility (Porta *et al.* 1999). Using this criterion, the accumulation of organic matter in marshy areas would take approximately 15 – 20 years to effectively mitigate the affects of an oil spill, once marshy vegetation was established in the affected area.

Table 3. Comparison of Total Hydrocarbon Concentration and Toxicity in the Oil Spill Adjacent to the Gas Processing Complex in La Venta (Tabasco, Mexico).

Sample	Toxicity Units[a]	Hydrocarbon Concentration (ppm)[a]
Algal mat (on top of oily crust)	12.84 (8.85 – 39.96)	118,675 (0,785 - 141,137)
Oily Crust	19.18 (9.17 – 35.97)	800,000 (740,000 – 840,000)
Contaminated soil, spill area	26.49 (20.40 – 38.89)	190,283 (69,241 - 283,102)
Mangrove leaf litter (uncontaminanted nearby area)	26.67 (15.39 – 45.21)	Not determined
Very low or non oil impacted soil (mangrove)	28.37 (11.57 – 57.81)	428 (0 – 1,703*)

(from Adams *et al.*, 1999)

[a]. Average (range of values).

Toxicity units = $1/EC_{50}$, EC_{50} (effective concentration-50), from the Microtox accute toxicity bioassay. The EC_{50} is the concentration of sample in a dilution which reduces biological activity (in this case bioluminescence) by 50 %.

*: only two of five samples had hydrocarbons (437 ppm and 1703 ppm).

Note that the toxicity of the oily crust is less than in the leaf litter from the uncontaminated area, and that the toxicity in the contaminated soil below the oily crust (contaminated soil, spill area) was in the same range as the very low or non-impacted soil.

Another technique which Pemex has used successfully was applied near the separation battery in the San Ramón Oil Field in western Tabasco State, Mexico. Here, fresh organic peat, collected from a nearby marshy area, was placed directly on top of oil impregnated soil, to form a cushion of new soil approximately 20 cm thick. Subsequently, cattails, ferns and some moisture resistant grasses were planted on top of the new material (Pérez, 1998). Eight years latter the area was sampled to evaluate soil toxicity (Adams and Ricalde, 2003). During the sampling of this site it was observed to be completely coverd with the planted vegetation plus other secondary herbaceous species as well as the swamp bush *Myrica serifera*. Upon taking core samples from this area, the division between and older oil stained material and the newer organic peat was clearly distinguishable. Samples were taken of both of these horizons as well as from surface horizons of uncontaminated native soil. The natural vegetation in this area is "tazistal" an asociation of a small spiny marsh palm, "taziste" (*Acoelorrhaphe wrightll*), and a sharp edged sege, "navajuela" (*Cyperus ferax*) (Magaña, 1995; West *et al.* 1985). The buried, oil stained soil was found to still present toxicity, but this toxicity was not present in the surface horizon. The surface soil conditions had been restored to be less toxic, and in the range of the soil in the control area (see Table 4).

Table 4. Toxicity in the Phytorremediation Treated Area in the San Ramón Oil Field.

Sample	Toxicity Unitis[a]
Control Area (tazistal)	22.2 (BQL* - 69.9)
Phytoremediation - superficial horizon (fresh organic material)	17.8 (13.1 – 50.5)
Phytoremediation - deep horizon (not treated)	43.5 (40.6 – 45.9)

(from Adams and Ricalde, 2003).
Toxicity units = 1/EC50, see notes Table 3.
a: Average (range of values).
*BQL: below quantification limits (< 10 TU in 10-6 ppm-1), see notes Table 3.
Note that the toxicity of the untreated deep horizon is roughly twice as high as the control area, while the surface horizon (treated) is in the same range as the control, and maybe even a little less toxic.

In another project, white and black mangrove (*Laguncularia racemosa* and *Avicenna germinans*) where used in an attempt to biologicaly confine very weathered surface spills (see Figure 4), in this case first cutting holes in the surface crust or in the surface of oil impregnated soil and then transplanting entire plants with there bulk of associated soil (Adams *et al.* 1999). Some of the most important factors in this project were the soil moisture in proportion to the hydrocarbon concentration, the heat generated in the oily crust, and the species of mangrove used.

In one of the experimental areas, the organic soil was contaminated with weathered oil up to a concentration of 19 % (dry weight basis), and covered with a very weathered crust of oily material (79 % hydrocarbons), but even under these extreme conditions, resistance and growth of several herbaceous species was observed. In this area the proliferation and invasion of the oil stained soil by cattails has also been observed. In another area with hydrocarbon concentrations up to 33 %, in an organic soil and without the oily crust, a diversity of marsh species have been found to grow well and completely cover the contaminated soil. However, at hydrocarbon concentrations of 45 – 50 % or more, the survival of planted or naturally occurring vegetation was almost nil, drying out completely during the dry season. It appears that at this concentration of hydrocarbons, the oil displaces the moisture in the soil, reducing field capacity (moisture retention capacity), and during the dry season there is not sufficient moisture to maintain marsh species.

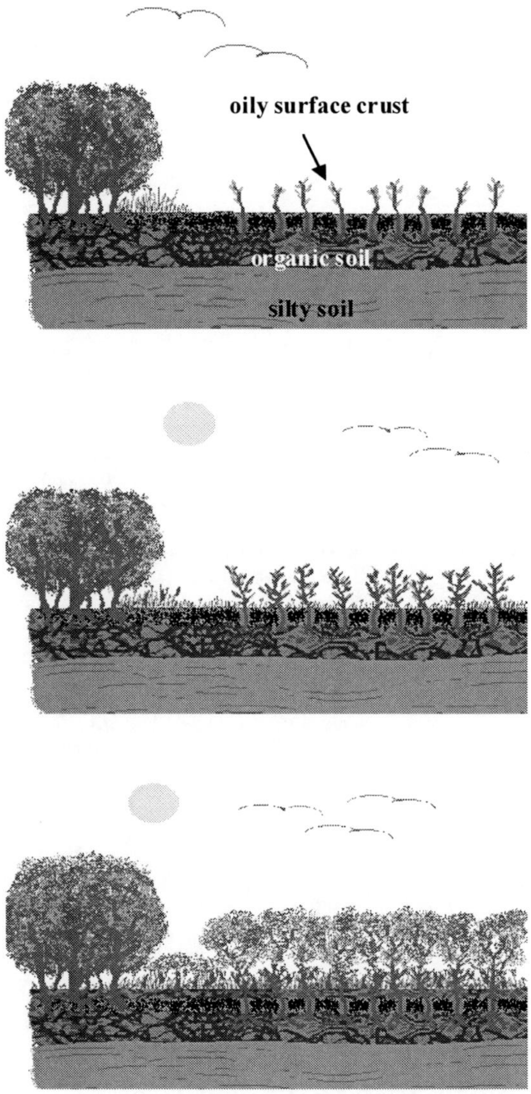

Phytoremediation Project Startup

A mangrove plantation is established in a weathered oily crust, cutting through the crust and sowing plantules with their attached, uncontaminated mineral soil, so that their roots can make contact with the moisture and nutrients in the underlaying soil. Note that at the edge of the oil spill, herbaceous and mangrove vegetation have already been established naturally and have started to cover the spill with leaf litter and fresh organic soil.

Project Advancement

The mangrove plants begin to root in and become established. They develop in diameter, height and follage. Secondary herbaceous species start to grow on the produced leaf litter.

Biological Confinement of the Spill

Over the years, a considerable amount of leaf litter is converted into organic soil, burying the spill and forming a layer of new, uncontaminated soil. Various herbaceous species are estab-lished under the trees. Also, these conditions allow for the natural seeding of mangrove and their subsequent survival.

Figure 4. Phytoremediation using Mangroves.

In the experimental area with the oily crust (79 % hydrocarbons), soil heating was observed below the crust which elevated the soil temperature 3 – 5 ° C higher than the background levels, reaching temperatures of more than 42 °C. At these temperatures it is very improbable that mangrove species will survive (Barth and Lieth 1982), however, there was greater success with black mangrove (*Avicenna germinans*) than with white mangrove (*Laguncularia racemosa*). During plant collection, a few individuals of black mangrove were accidentally collected among the hundreds of white mangrove plants, and later planted along with the white mangroves. The survival rate of white mangrove after three years was less than one per cent, but 100 % of the black mangrove plants had survived and showed vigorous growth (see Figure 5; Adams and Castillo, 2001). Although the number of individuals of black mangroves planted was very low and makes any statistical evaluation difficult, it is likely that this species is indeed more resistant. Its physical niche in the landscape is in natural depressions between the shorelines of water bodies (such as coastal rivers or lagoons) and higher, inland areas. In these depressions, when the overall water level drops (such as in the dry season), small shallow ponds are formed in which salty water stagnates and heats up, slowly evaporating, and increasing the salinity and temperature (Lugo and Snedeker, 1974; Rützler and Feller, 1996).

Table 5. Hydrocarbon Concentration in Organic Soil vs. Vegetative Development.

Hydrocarbon concentration (%)	Oily crust	Vegetation type	Observations
19 %	Yes (79 % oil)	Cattail (*Typha latifolia*)	Invasion of the area by vegetative reproduction and root extension below the oily crust; observations from La Venta, Tabasco, México.
33 %	No	Secondary marsh vegetation: "molinillo" (*Cyperus gigantus*), cattail (*Typha* sp.), popay (*Thalia geniculata*), ferns, grasses, various *Cyperus* spp. and the marsh bush *Myrica serifera*	Relatively high species diversity, plant height of approx. 2 ½ m, complete vegetative cover, vegetation does not appear to be excessively stressed during the dry season; observations from the Sánchez Magallanes Oil Field, Tabasco, Mexico.
45 – 50 % or more	No	Almost none	During the rainy season and "northerns" season (light rain, windy weather) survival of planted white mangrove is high (73 %), but during the dry season all planted individuals die off. Natural vegetation is non-existent; observations from the Sánchez Magallanes Oil Field, Tabasco, Mexico.

Figure 5. Resistance of Black Mangrove for use in Phytoremediation.

Figure 6. Experimental Phytoremediation Plantation in the Sanchez Magallanes Oil Field.

PROSPECTIVES

Each kind of ecosystems has its own particular characteristics which help mitigate the impact caused by oil spills. In areas with good drainage, the root zone biodegradation, i.e. mineralization and humification, are very important processes, however, in marshy areas these aerobic processes are of very limited importance. In marshy areas (marshes, swamps, floodable pastures) the principal processes for natural attenuation are: 1) the strong absorption of medium and heavy range hydrocarbons to the abundant soil organic matter, and 2) the biological confinement (interment) of oil spills due to the growth of marshy vegetation on top of the spill and the subsequent production of leaf litter and organic soil. These processes are very specific for floodable soils.

It is important to mention that these mechanisms of natural recovery, although they do not eliminate the contaminants from the soil, do effectively maintain such compounds sequestered, with the formation of an organic barrier between the potentially harmful contaminated material and possible ecological receptors (such as wildlife, for example). In the long run, these processes make it possible to meet qualitative environmental cleanup criteria. For example, in Mexican environmental legislation, in the General Law of Ecological Equilibrium and Environmental Protection, abbreviated as LGEEPA according to its name in Spanish, (Ley General de Equilibrio Ecológico y Protección al Ambiente, SEMARNAP 1997) the following qualitative criteria are mentioned:

"Those residues which accumulate or which could accumulate, and which are deposited or infiltrate into the soil should meet the conditions necessary to prevent or avoid:

I – the contamination of the soil;
II – harmful alterations to the biological processes in the soil;
III – alterations in the soil which impair its use or exploitation;
IV – risks and health problems"
(art. 136)

Once a spill is completely confined biologically, and a layer of new organic soil is formed which provides fertility to the vegetation (approx. 30 cm deep), practically all harmful alterations to the biological processes in the soil and is problems in its ability to be used, are avoided.

This second point (soil use) is very important with respect to the biological confinement of oil spills and its legal implications. In this same chapter of the law (LGEEPA), in article 134, it specifies that it is necessary to remediate a site so that is can be use according to its natural vocation (according to a soil use study). If a site is apt to be used as marsh and wildlife refuge (say, because of its hydrology), it is not necessary to remediate it so that is may be used for some other purpose, such as cattle ranching, residential, etc. With respect to the water in a marsh, the small amount of hydrocarbons that could remain in the water are levels which are not toxic, and in a level which is accepted by the National Water Commission (Adams, 2000). Likewise, it is not necessary to protect the organoleptic properties of the water in marsh areas, since the water is naturally of poor quality (turbidity, fetid odors, sediments, salinity) and not amenable to being potable or used for human consumption.

Another factor which will be very important for the use of natural attenuation or phytoremediation in the future is social acceptance. Although it is possible to convince industries and possibly environmental authorities of the applicability of these methods (using appropriate documentation of the process), it is also necessary that the community accept these methods and does not insist on more intensive remediation methods. In some areas (for example, in southern Veracruz State, Mexico) contaminated sites are not so political and false claims made to blackmail the petroleum industry are minimal. In such areas, natural attenuation and phytoremediation programs could be implemented if reasonable compensations were also paid to affected parties. In other areas, especially in parts of Tabasco State (Mexico), this would be a serious social challenge. In such places, it would be more advisable to implement superficial *in situ* remediation projects (chemical or bioremediation) as a pre-treatment to phytoremediation.

Which ever way that these methods are used (directly or in combination with other technologies), their application in tropical and semitropical environments can result in a considerable cost reduction in remediation projects, and they are especially attractive alternatives for sites where the low (or null) economic gains to be obtained from the site don't justify the use of intensive technologies. Besides this, natural attenuation and phytoremediation have their own virtues (Lin and Mendelssohn 1998). Unlike the majority of *in situ* technologies, and unlike all *ex situ* technologies, natural attenuation and phytoremediation do no chemically or physically damage the soil. In many remediation project using intensive technologies, the physical or chemical damage to the site can be considerable and sometimes even greater than that caused by the contaminants. This is true especially when the project begins by removing all of the topsoil (and thereby the fertility) as well as the surface vegetation from the contaminated area. Also, the application of some kinds of chemical used in more intensive remediation methods (for example soil washing, solvent extraction, use of quicklime, and even over fertilizing) may result in more toxic conditions to the soil biota than the contaminants themselves. However, when using remediation methods of low intensity, such as those presented in this paper, increase the *ecological value* of the site, favoring the formation of new soil, fertility, vegetative cover, and with time, increasing primary productivity and biodiversity in marshy areas.

REFERENCES

[1] Adams S., R. H., (2000). Estudio de Riesgo Toxicológico In: Caracterización de Suelo, Subsuelo y Mantos Freáticos Contaminados en el CPG, La Venta. Corporación en Ingeniería Ambiental S. A. de C. V. Unpublished report. Corp. Ing. Ambiental – Pemex Gas y Petroquímica Básica.

[2] Adams S., R. H. and Castillo, A. O., (2000). Fitorremediación y Fitomitigación para Restauración de Localizaciones Petroleras. Kuxkulkab, 5 (11), 30-34.

[3] Adams S., R. H. and Castillo A., O. (1998). Fitorremediación con Mangle Blanco de Áreas Impactadas por Hidrocarburos y su Manejo como Agrosilvo-ecosistema: Parte I. Kuxkulkab, 5-6, 32 – 42,

[4] Adams S., R. H. and Ricalde Z., S. C. (2003). Toxicidad en Suelos del Activo de Producción Petrolera Cinco Presidentes, Veracruz y Tabasco. Congreso Internacional de Ingeniería Ambiental. Instituto Tecnológico de Minatitlán, 10 – 15 November, Minatitlán, Veracruz. .

[5] Adams S., R.H. and Castillo A., O. (2001). Fitorremediación de Áreas Afectadas por Derrames de Hidrocarburos Usando Especies de Mangle. XV Congreso Mexicano de Botánica. Sociedad Botánica de México, A.C. 16 -18 December, Querétaro, Querétaro.

[6] Adams S., R.H., Castillo A., O., Zavala C., J. and Palma-López, D. J. (1999). Recuperación con Mangle Blanco (Laguncularia racemosa) de Áreas Impactadas por Hidrocarburos y su Manejo como Agrosilvo-ecosistema en la Zona Costera de Huimanguillo y Cárdenas, Tabasco. Final report, Proyecto MO76, Comisión Nacional para el Concocimiento y Uso de la Biodiversidad (CONABIO), The John T. and Catherine D. MacArthur Foundation –) – Universidad Juárez Autónoma de Tabasco, Villahermosa, Tabasco.

[7] Al-Assi, A.A. (1993). Uptake of poliaromatic hydrocarbons by alfalfa and fescue. Masters of Science Thesis. Kansas State University, Manhattan, Kansas

[8] Anderson, T. A. U. and Coats, J. R. (1994). Bioremediation through Rhizosphere Technology. *American Chemical Society*, Washington, D.C.

[9] Aprill, W. and Sims, R. C. (1990). Evaluation of the use of prairie grasses for stimulating polycyclic aromatic hydrocarbons treatment in soil. *Chemosphere,* 20 (1-2), 253-265.

[10] Atlas, R. M. and Bartha R. (1987) Microbial ecology, fundamentals and applications (2d ed). The Benjamin/Cummins Publishing Company, Inc. Menlo Park, California.

[11] Barth H., and Lieth, H. (1982) Applicability of Mangroves for the Development of Ecologically Based Mariculture Systems in Subtropical and Tropical Coastal Deserts. Studies on Aquatic Vascular Plants. (ed. J. J. Symoens, S. S. Hooper and P. Compere), pp. 235 - 239. Brussels.

[12] Bolaynas L., R. (2003). Elaboración de criterios para la limpieza de sitios contaminados por hidrocarburos en el estado de Tabasco y Norte de Chiapas. Bachelor's Thesis, Universidad Juárez Autónoma de Tabasco. Villahermosa, Tabasco.

[13] Briggs, G. G., Bromilow, R. H., and Evans, A. A. (1982) Relationships Between Lipophilicity and Root Uptake and Translocation of Non ionized Chemicals by Barley. Pest. Sci., 13 (5), 495-504.

[14] Burken, J. G. and Schnoor, J. L. (1997) Predictive relationships for uptake of organic contaminants in hybrid poplar trees. *Environ. Sci. Technol.*, 32 (21), 3379 – 3385.

[15] Carman, E. P. (1997) Using phytoremediation to address fuel oil contaminated soil at an active industrial facility - a low cost, non-invasive, in-situ alternative. International Business Communications, Second Annual Conference on Phytoremediation, 18 – 19 June. Seattle, Washington.

[16] Carman, E.P., Crossman, T. L. and Gatliff, E. G. (1997) Phytoremediation of Fuel Oil Contaminated Soil. International Business Communications Second Annual Conference on Phytoremediation, 18 – 19 June. Seattle, Washington.

[17] Chávez, M. L. (1996) Personal communication about phytoremediation projects at Costa Rica. Refinación Costarricense de Petróleo, San José, Costa Rica.

[18] Clark, J. R. (1996) Coastal Zone Management Handbook. Lewis Publishers, Boca Raton, Florida.

[19] Dickinson, N.M., Turner, A.P. and Lepp N.W. (1991) How do tress and other long-lived plants survive in polluted environments? *Funct. Ecol.*, 5, 5-11.

[20] Drake, E. N. (1997) Industry Acceptance - Assessment of current prospects and limitations for industrial applications. International Business Communications Second Annual Conference on Phytoremediation,. 18 – 19 June Seattle, Washington.

[21] Drake, E. N. (1997a). Phytoremediation of aged petroleum hydrocarbons in soil. International Business Communications Second Annual Conference on Phytoremediation, 18 – 19 June. Seattle, Washington.

[22] Duke, N.C.; Pinzón M., Z. S. and Prada T., M. C. (1997) Large-scale damage to mangrove forest following two large oil spills in Panama. Biotropica ,29 (1), 2-14

[23] Edwards, D.A., Andriot, M.D., Amoruso, M.A., Tummey. A.C., Bevan, C.J., Tveit, A., Hayes, L.A., Youngren, S.H., and Nakles, D.V. (1997) Development of Fraction Specific Reference Doses (RfDs) and Reference Concentrations (RfCs) for Total Petroleum Hydrocarbons (TPH), Volume 4. Total Petroleum Hydrocarbon Criteria Working Groups Series. The Association of Environmental Health and Sciences, Massachusetts, U.S.A. http://www.aehs.com/publications/catalog/contents/tph.htm

[24] Erickson, L. E., Banks, M. K., Davis, L. C., Schwab, A. P., Muralidharan, N., Reiley, K. and Tracy, J.C. (1993). Using vegetation to enhance in situ bioremediation. http://www.engg.ksu/HSRC/phytorem.

[25] Escalante-Espinosa E., Gallegos-Martínez, M. E., Favela-Torres, E. and Gutiérrez-Rojas, M. (2005) Improvement of the hydrocarbon phytoremediation rate by Cyperus

laxus Lam. inoculated with a microbial consortium in a model system. *Chemosphere*, 59, 405-413.

[26] Gallegos-Martínez, M., Gómez Santos, A., González Cruz, L., Montes de Oca García, M.A., Yáñez Trujillo, L., Zermeño Eguía Lis, J.A. and Gutiérrez-Rojas, M. (2000) Diagnostic and resulting approaches to restore petroleum-contaminated soil in a Mexican tropical swamp. *Water Sci. Technol.,* 42 (5,6), 377-384.

[27] Glooshenko, W. (1996). Ecología de humedales. In: Manejo de Ecosistemas Tropicales. (ed O. Castillo.). Antología para el Diplomado de Manejo de Ecosistemas Tropicales. Universidad Juárez Autónoma de Tabasco. Villahermosa, Tabasco.

[28] Grey, N. (1994) Personal communication of experiments using native grasses for the recovery of hydrocarbons impacted soils in Canadian prairies. ICI - Zeneca, Ltd. Toronto, Ontario, Canada.

[29] Gustafson, J., Griffith Tell, J. and Orem, D. (1997) Selection of Representative TPH Fractions Based on Fate and Transport Considerations, Volume 3. Total Petroleum Hydrocarbon Criteria Working Groups Series. The Association of Environmental Health and Sciences. Massachusetts, U.S.A. http://www.aehs.com/publications/catalog/contents/tph.htm

[30] Hinchmann, R. R., Negri, M. C. and Gatliff, E. G. (1997) Phytoremediation: Using Green Plants to Clean Up Contaminated Soil, Groundwater, and Wastewater. International Business Communications Second Annual Conference on Phytoremediation, 18 – 19 June. Seattle, Washington,.

[31] Jackson, L. (1997) Why choose phytoremediation - Looking through the eyes of the customer. International Business Communications Second Annual Conference on Phytoremediation, 18 – 19 June. Seattle, Washington.

[32] Lewis, R. R. (1990) Creation and Restoration of Coastal Wetlands in Puerto Rico and the U.S. Virgin Islands. In: Wetland Creation and Restoration, the Status of the Science. (ed. J. A. Kusler y M. E. Kentula,.). Island Press, Washington, D.C.

[33] Lin, Q. and Mendelssohn, I. A. (1998) Phytoremediation for oil spill cleanup and habitat restoration in Louisiana coastal marshes: Effects of marsh plant species and fertilizer. Louisiana Applied Oil Spill Research and Development Program (OSRADP), Technical Report Series 97-006. Louisiana State University/Louisiana Oil Spill Coordinator-Office of the Governor, Baton Rouge.

[34] Lugo, A. E., and Snedeker, S. C. (1974) The Ecology of Mangroves. Annual Review of Ecology and Systematics, 5: 39-64.

[35] Magaña, M. (1995) Catálogo de Nombres Vulgares y Científicos de Plantas de Tabasco. Universidad Juárez Autónoma de Tabasco, Tabasco, México.

[36] Mendelssohn, I. A., Hester, M. W., and Pahl J. W. (1996) Environmental effects and effectiveness of in situ burning in wetlands; considerations for oil spill cleanup. Louisiana Applied Oil Spill Research and Development Program (OSRADP), Technical Report Series 95-010. Louisiana State University/Louisiana Oil Spill Coordinator-Office of the Governor, Baton Rouge.

[37] Merkl, N., Schultze-Kraft, R. and Infante, C. (2004) Phytoremediation in the Tropics – The effect of crude oil on the growth of tropical plants. *Bioremediation J.,* 8 (3-4), 177-184.

[38] Moreno C., E. (1999) Influencia de la Vegetación en Histosoles de los Humedales de la Cuenca Baja del Río Tonalá en Tabasco. Bachelor's Thesis, Universidad. Juárez Autónoma de Tabasco, Villahermosa, Tabasco.

[39] Olán, C.D. (2000) Estimación de Niveles de Hidrocarburos Totales de Petróleo, Poliaromaticos Carcinogénicos y Metales, en el Activo Cinco Presidentes Tabasco. Bachelor's Thesis, Universidad. Juárez Autónoma de Tabasco, Villahermosa, Tabasco.

[40] Pérez, J. (1998). Personal comunication. Pemex - Seguridad Industrial y Protección Ambiental, Pemex, Activo Cinco Presidentes.

[41] Porta C., J., López-Acevedo R., M. and Roquero d. I.B., C. (1999). Edafología para la Agricultura y el Medio Ambiente. Ediciones Mundi-Prensa, Madrid.

[42] Rodríguez F., A. (1996). Personal communication. Texas A & M University, Corpus Christi.

[43] Rützler, K., and Feller, I. C. (1996) Caribbean Mangrove Swamps. Scientific American, March.

[44] Salt, D. E., Smith, R.D. and Raskin, I. (1998) Phytoremediation. Ann. Rev. Plant Physiol. *Plant Mol. Biol.*, 49, 643-668.

[45] Schnoor, J. L., Licht, L. A., McCutcheon, S. C., Wolfe, N. L., and Carreira, L. H. (1995) Phytoremediation of Organic and Nutrient Contaminants. *Environ. Sci. Technol.* 29 (7), 318 – 323.

[46] Schwab, A. P. 1997) Phytoremediation of Diesel-Contaminated Soil at the U.S. Navy's Craney Island Fuel Facility – Enhancing Degradation. International Business Communications Second Annual Conference on Phytoremediation. Seattle, Washington.

[47] Secretaría de Medio Ambiente, Recursos Naturales y Pesca. 1997. Ley General de Equilibrio Ecológico y Protección al Ambiente. México, D.F.

[48] Singer, A.C., Crowley, D.E. and Thompson, I.P. (2003) Secondary plant metabolites in phytoremediation and biotransformation. *Trends Biotechnol.*, 21 (3), 123-130.

[49] Van Fleet, E. (1994) Tampa Bay Spill: Tracking the fate of Oil. Proceedings of: Gulf of Mexico and Caribbean Oil Spills in Coastal Ecosystems: Assessing Effects, Natural Recovery, and Progress in Remediation Research. U.S. Dept. Interior, New Orleans, Louisiana, USA.

[50] Vorhees, D., Gustafson, J., and Weisman, W. (1997). Human Health Risk-Based Evaluation of Petroleum Contaminated Sites: Implementation of the Working Group Approach Volume 5. Total Petroleum Hydrocarbon Criteria Working Groups Series (TPHCWG). The Association of Environmental Health and Sciences, Massachussettes, U.S.A. http://www.aehs.com/publications/catalog/contents/tph.htm

[51] West, R; N. Psuty; and B. Thom. 1985. Las Tierras Bajas de Tabasco Gobierno del Estado de Tabasco. Instituto de Cultura de Tabasco. Villahermosa Tabasco.

[52] White, Jr., P.M., Kirkpatrick, W.D., Thoma, G.J. and Wolf, D.C., 2000. Using plants to remediate crude oil-contaminated soil. Proceedings of The 7th International Petroleum Environmental Conference, Albuquerque, Nuevo México.

[53] Woodward, D., and Day, M. 1996. Intrinsic Bioremediation. PEMEX Seminar, Remedial Project Management. Pemex Exploración y Producción, Seguridad Industrial y Protección Ambiental, Región Sur. Villahermosa, Tabasco.

[54] Wyoming Department of Environmental Quality. 2003. Fact Sheet No.12, Soil Cleanup Level Look-Up Table Under the Voluntary Remediation Program, 05/09/03

v.5, Appendix, Cleanup Levels for Total Petroleum Hydrocarbons in Soil and Groundwater.

[55] Zavala C., J., Gavi R., F., Adams S., R. H., Ferrera-Cerrato, R., and Palma-López, D. J. 2002. Hidrocarburos del Petróleo y Tecnologías de Biorremediación para Suelos de Tabasco. En: Plan de Uso Sustentable l Suelos de Tabasco, Vol. II. Colegio de Posgraduados-ISPROTAB. pp 125 – 156.

ACKNOWLEDGMENTS

We would like to thank Mr. Juan Pérez de la Cruz, formerly of Seguridad Industrial y Protección Ambiental (SIPA) in Pemex Exploración y Producción in Agua Dulce, Veracruz for sharing his experiences about preliminary restoration projects in the region. Also, to Dr. Joel G. Burken of the University of Missouri-Rolla, for his valuable information with respect to transpiration and contaminant behaviour in roots.

We would also like to thank Mr. Sergio Morales and Mr. J. Abisenas Alvarez Rivera for their participation in environmental diagnostic and phytoremediation projects, as well as Mrs. Verónica Domínguez Rodríguez for her numerous suggestions and detailed revision of this paper.

Finally, we would like to thank the Comisión Nacional para el Conocimiento y Uso de la Biodiversidad (CONABIO) and The John T. and Catherine D. MacArthur Foundation for financial assistance.

In: Remediation of Soils and Aquifers
Editors: Luis G. Torres and Erick R. Bandala
ISBN 978-1-60741-372-1

Chapter 2

SOIL WASHING WASTEWATER TREATMENT USING ADVANCED OXIDATION PROCESSES

Erick R. Bandala **and Luis G. Torres***
UDLA Puebla and UPIBI-IPN

ABSTRACT

A review of the state-of-the-art on the available technologies for soil washing wastewater treatment is provided. The application of the surfactant enhanced soil washing (SESW) technology for the site restoration involving soil contaminated with hydrocarbons and pesticides is briefly described and the analysis of the problem associated with the wastewater generated during the SESW processes is described. From the available literature reports, the best approaches to the wastewater treatment are analyzed and a perspective for further scientific and technological challenges is described.

INTRODUCTION

Soil washing technology involves the use of solvents and the presence of surfactants to increase the solubility of non-aquous phase liquids (NAPLs). The washing mixture is then injected into the soil were it extracts adsorbed contaminants. Following the standard procedure, the surfactant solution, including the extracted contaminants, is recovered from the soil. This solution, containing the added surfactant and the extracted pollutants, is usually high toxic and include very hazardous materials that may jeopardize the environment if washing wastewater is released without the proper effective treatment. If the effect of the generation of potentially hazardous wastewater is considered, it is clear that the surfactant extraction technology must involve the further capture and/or removal of the pollutants included within.

* Corresponding author: erick.bandala@udlap.mx

WASTEWATER TREATMENT TECHNOLOGIES

Wastewater resulting from surfactant enhanced soil washing (SESW) processes may be a very complex matrix and include a wide variety of NAPLs related pollutants such as metals (Torres et al., 2007b; Lambert et al., 2004; Torres et al., 2005), hydrocarbons (Torres et al., 2007b; Torres et al., 2005; Torres et al., 2004; Iturbe et al., 2004a,b; Lopez et al., 2004; Torres et al., 2007a), pesticides (Aguilar, 2008; Lee et al., 2004; Mata et al., 2002) and surfactants (Bandala et al., 2008a,b).

The main problem of the treatment of soil washing wastewaters is related with the complexity of the matrix and, as stated earlier, the presence of the potentially toxic contaminants extracted from the soil. The usual procedure is to submit the wastewater through different physical-chemical and/or biological processes with the goal of remove the pollutants and reuse the water in the washing process. The processes currently involved frequently consist of coagulation-flocculation-sedimentation steps coupled with, for example, an activated sludge system. Additionally, some others procedures used to separate the different components of the wastewater are the use of ultra-filtration (Sabatini et al., 1998), interchange columns (Cheng and Sabatini, 2002) or biodegradation (Torres et al., 2007; Scott and Jones, 2000), among others. The use of mass transference process as ultra-filtration or interchange columns, posses one very important disadvantage: the generation of a diluted phase with low pollutant concentration and a concentrated phase, usually high toxic and including the potentially hazardous chemicals used in the remediation process and those extracted from the contaminated site (Yuan and Marshall, 2007). The application of biodegradation, on the other hand, becomes more complicate because the presence of high toxic and persistent pollutants in the effluent that may inhibit the efficiency of the microorganism strains and completely avoid biological treatment (Sarria et al., 2003).

It is obvious that, considering the lack of appropriate treatment technologies for application in soil washing wastewater, the development of alternative treatment methodologies is a very interesting scientific and technological issue that must be accomplished before the complete adoption of SESW at full scale for soil restoration in contaminated sites.

ADVANCED OXIDATION PROCESSES (AOPS)

Among the emerging methodologies proposed as alternative treatment to wastewater including the presence of highly toxic and persistent pollutants, Advanced Oxidation Processes (AOPs) are considered a highly efficient one. AOPs have been described as a good alternative to eliminate persistent and toxic pollutants from water when conventional treatment processes are not efficient (Bandala et al., 2008c). Theses processes are based on the generation, using both photochemical and non-photochemical processes, of strong oxidizing species able to react with any organic or inorganic compound and have been used as alternative processes for water treatment applications with pesticides, dyes, surfactants, hydrocarbons, toxins and a wide variety of organic pollutants (Chacon et al., 2006; Bandala et al., 2007a,b; Gelover et al., 2000; Bandala et al., 2008a,b,c; Bandala et al., 2004; Bandala and Estrada, 2007).

Among AOPs, Fenton and Fenton-like processes are highly oxidant methodologies which are considered as very interesting alternative for its application to complex wastewater matrix (Anipsitakis and Dionysiou, 2003;2004). The Fenton method is a common oxidation process that includes the formation of hydroxyl radicals (oxidation potential 2.8 V) by the combination of ferrous salt and hydrogen peroxide. The reaction starts with the combination of H_2O_2 with Fe^{2+} or Fe^{3+} as shown in eq. (1) and (2) (Orozco et al., 2008):

$$Fe^{2+} + H_2O_2 \rightarrow Fe^{3+} + HO^- + HO^\bullet \tag{1}$$

$$Fe^{3+} + H_2O_2 \rightarrow Fe - OOH^{2+} + H^+ \tag{2}$$

The addition of radiation, like UV and visible light, accelerates the generation of hydroxyl radicals and the Fe^{3+} generated by the Fenton's reaction, as proposed in (1) is continuously reduced to Fe^{2+} in a reaction known as photo-Fenton process and the reaction starts again using the photo-generated Fe^{2+} as proposed in eq. (3) (Orozco et al., 2008; Chacon et al., 2006):

$$FeOH^{2+} + h\nu \rightarrow Fe^{2+} + HO^\bullet \tag{3}$$

The use of other transition metals with the combination of oxidants had being studied (Anipsitakis and Dionysiou, 2004). For example, the use of cobalt as the transition metal instead of iron and peroxymonosulfate as the oxidant in replace of hydrogen peroxide has been tested as a very efficient Fenton-like process (Anipsitakis and Dionysiou, 2003;2004; Bandala et al., 2007a,b; Bandala et al., 2008a,b). The cobalt/peroxymonosulfate (Co/PMS) system have been also tested for the degradation of several organic pollutants (Anipsitakis and Dionysiou, 2003;2004; Bandala et al., 2008a,b). It generates sulfate radicals are highly reactive as shown in eq. (4) (Anipsitakis and Dionysiou, 2004) :

$$Co^{2+} + HSO_5^- \rightarrow Co^{3+} + SO_4^{\bullet -} + HO^- \tag{4}$$

The combination of UV light (using lamps or solar radiation) to promote the formation of radical species in this system was recently explored with interesting result for the application of this technology to water and wastewater treatment for the removal of highly toxic and persistent pollutants (Bandala et al., 2007a,b; Bandala et al., 2008a,b).

Hydrocarbon Containing SESW Wastewater Treatment by AOPs

Application of photo-assisted oxidation processes has demonstrated high possibilities for soil washing wastewater treatment. Chu et al. (2005) demonstrated that the utilization of a soil washing/photodegradation process is a promising strategy for site remediation. They found, that, despite previous reports about humic substances presents in soil washing wastewater are capable to inhibit free radical chain reactions (Zepp et al., 1975), humic materials plays an important role in the photolytic water detoxification process agreeing with other recent reports (Chan and Chu, 2005). Other interesting approach for the AOPs application to the elimination of pollutants from soil washing wastewater are, for example, the use of photo-Fenton process for the treatment of the effluent (Riaza et al., 2007).

Riaza et al. (2007) found an efficiency of 87 and 100% for desorption and oxidation processes, respectively. The use of the dark Fenton reaction (Saxe et al., 2000) have also provided interesting results in the removal of PAHs in surfactant enhanced soil washing effluents. Other hydroxyl radical based advanced oxidation (HR-AOP) process tested so far for soil washing wastewater treatment is the widely known TiO_2 photocatalyzed process (Maillacherovu et al., 2001). In this interesting work, authors propose the use of UV assisted TiO_2 photocatalytic degradation of the surfactants in the soil washing wastewater followed from the biological treatment of the generated effluent. They found high improvements to the biodegradability of the effluent by the use of the photocatalytic process.

In a recent work, we have reported the application of photo-assisted Fenton-like processes for surfactant degradation (Bandala et al., 2008b). In this work, we have successfully complete degrade the surfactant used for soil washing procedures, sodium dodecyl sulfate (SDS), by the use of both, photo-Fenton and Co/PMS/UV using solar radiation as driven force of the processes. The concentration of SDS tested in our work was as high as 0.1 % and the tests were performed in a bench scale system able to treat up to 10 L of water.

Figure 1 show a comparison among the photo-assisted and non photo-assisted processes for the Fenton reaction as function of reaction time. As shown, no decrease in the SDS initial concentration (0.1%) was determined by the use of hydrogen peroxide (32 mM) in dark experiments. By exposing the reaction mixture to sunlight, little SDS removal was detected (about 10%) after 60 minutes of irradiation. When using the Fenton reagents in the dark ($[Fe^{2+}]$=10 mM; $[H_2O_2]$=32 mM), achieved SDS degradation improved considerably to reach up to 60%.

Finally, to show the importance of including solar radiation in the system, when the Fenton reagents were submitted to solar radiation, the photo-Fenton process, almost complete SDS removal (total degradation about 95%) was obtained in only 30 minutes of irradiation.

For comparative purpose, figure 2 show the behavior of SDS when treated using the Co/PMS, under different experimental conditions, as function of reaction time. No SDS degradation was determined using dark PMS (32 mM) after 60 min of reaction. When solar radiation was used, under the same reaction conditions, SDS removal improved by the use of the oxidant agent alone to achieve 25%. Application of complete Co/PMS ($[Co^{2+}]$=10 mM; [PMS]= 32 mM) system in the dark improved the process reaching 30% of SDS removal in only 30 minutes. As shown in figure 2, no further increase in the SDS removal value was determined for the dark Co/PMS system after the first 30 minutes of reaction probably due to the consumption of the Co^{2+} in the reaction to generate Co^{3+} which is not able to activate the PMS oxidant (Anipsitakis and Dionysiou, 2004) producing the end of the oxidation reaction. It is very interesting to note that under the best conditions tested (10 mM of cobalt, 32 mM of PMS and solar radiation) complete SDS removal was achieved by only 5 minutes of solar radiation.

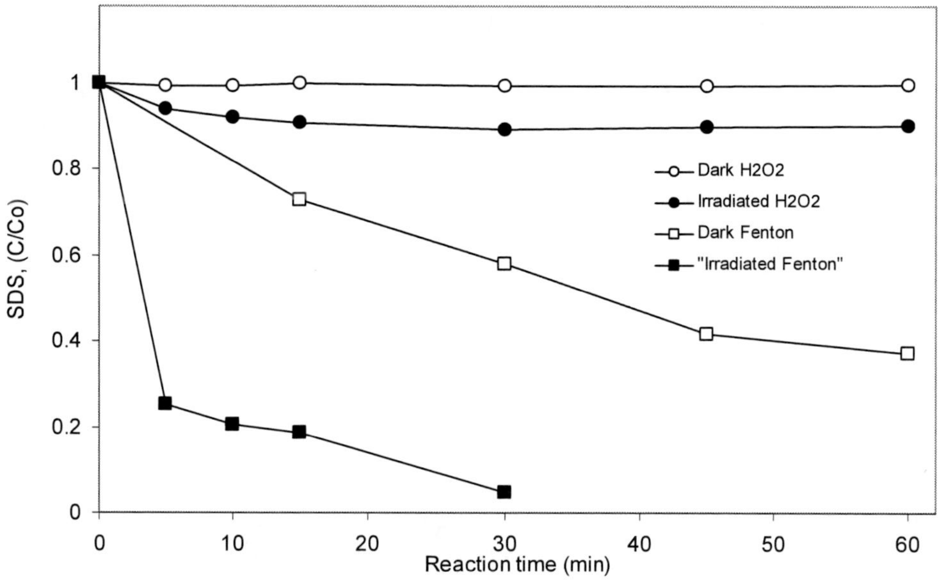

Figure 1. Comparison among different photo-assisted and non photo-assisted Fenton processes used for SDS degradation. The experimental conditions tested were $[Fe^{2+}]$= 10 mM; $[H_2O_2]$= 32 mM and average solar radiation (when apply): 920 W/m^2.

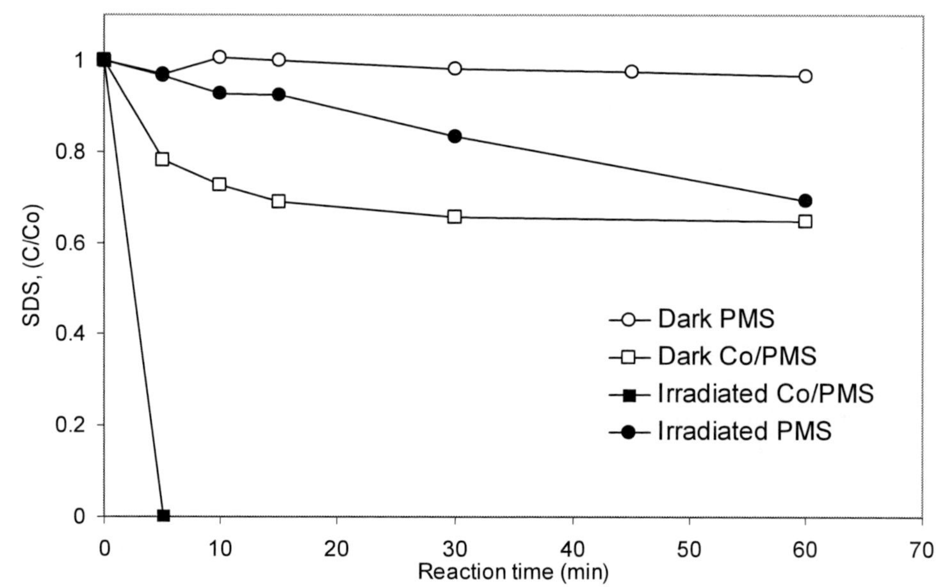

Figure 2. Comparison among different photo-assisted and non photo-assisted Co/PMS processes used for SDS degradation. The experimental conditions tested were $[Co^{2+}]$= 10 mM; [PMS]= 32 mM and average solar radiation (when apply): 930 W/m^2.

Comparison among the two tested processes allowed determining that Co/PMS seems to be more efficient to achieve SDS degradation under solar radiation conditions. The application of artificial UV radiation, generated using a UV lamp, showed that under these conditions Co/PMS was also able to achieve the highest SDS degradation and that this trend was observed independently or the radiation source.

In an even more recent report (Bandala et al., 2008b) the same processes were applied to real soil wastewater obtained from the treatment of polluted soil from the southeast of Mexico. In this work, soil washing wastewater produced from the washing of soil contaminated with petroleum hydrocarbon was submitted to the previously described Fenton-like processes to remove SDS and the organic matter, determined as chemical oxygen demand (COD).

The contaminated soil was obtained from Tabasco, Mexico and its characterization showed an average TPH content about 34.5-51.9 g/kg, 126 mS/cm of electrical conductivity and about 5.5-5.7 of pH (Torres et al., 2007b). The soil washing wastewater obtained from this soil has an SDS concentration of 3.7 g/L, 94.5 mg/L of oil and grease and 20.15 g/L of COD (Bandala et al., 2008b). By the use of the Co/PMS/UV system, we were able to complete degrade the SDS in the wastewater after 15 min of solar irradiation using 10 mM of cobalt and 32 mM of PMS and 897 W/m^2 of average solar radiation.

Figure 3 show the results of COD removal in the soil washing wastewater after treatment with both photo-assisted processes. It is worthy to note that, despite the Co/PMS/UV was the most efficient process for SDS degradation, photo-Fenton reaction achieved the highest COD removal (69%) in the same reaction time (60 min) compared with 51% achieved by the Co/PMS/UV system. We think that sulfate radicals generation in the Co/PMS/UV process occurs fast and propagation reaction produced by these radicals could degrade SDS as fast as observed. Nevertheless, the capability of photo-Fenton process to use both, UV and visible radiation results in the highest COD removal as showed in figure 3.

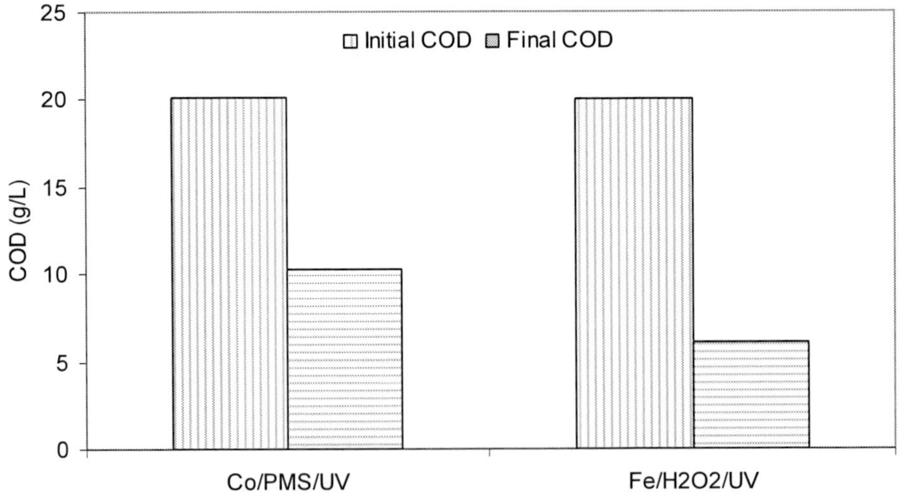

Figure 3. COD removal in the SESW wastewater using the two photo-assisted Fenton-like processes proposed in this work, Co/PMS/UV and photo-Fenton reaction.

PESTICIDE CONTAINING SESW WASTEWATER TREATMENT BY AOPS

Another interesting application of the soil washing process is the removal of pesticides from contaminated soil. Presence of pesticides always offers challenges to almost any technological approach applicable to the restoration of contaminated soils. Unlike oil derivate products, pesticides shows a wide variety of molecular structures which posses a very variable environmental behavior (Mata el al., 2002). We have tested the removal efficiency of the 2,4-dichlorophenoxyacetic acid (2,4-D), an herbicide widely used in Mexico (Bandala et al., 2007b; Santacruz et al., 2005). Two surfactants, anionic and non-ionic (SDS and Brij30, respectively), were tested at different concentrations for the extraction of the pollutant from soil spiked with 20 mg/kg of pesticide (Bandala et al., 2009). We found that, under the best conditions, 0.5% of SDS was capable to remove up to about 30% of the total pesticide during the first extraction and that the extraction efficiency improved up to 65% when two SDS extractions were carried out. Wastewater generated after the pesticide extraction (after two extractions) was characterized by HPLC and found that the actual 2,4-D concentration in the mixture was around 13 mg/L and the SDS concentration was 845 mg/L.

As showed earlier, SDS degradation is easy to carry out by the use of solar driven Fenton-like processes. We have been able to complete degrade SDS concentrations as high as 1000 mg/L within 5 minutes of irradiation when using the Co/PMS/UV process. It is clear that lower SDS concentration will be easily completely degraded under the tested conditions previously described.

Degradation of 2,4-D using solar driven Fenton-like processes have also been tested by our research group in the past (Bandala et al., 2007b). For the case of this pesticide, both photo-Fenton and Co/PMS/UV processes were tested under dark and irradiated conditions. Table 1 shows the reaction rate (*k*) values obtained for the non photo-assisted AOPs fitted to a first order kinetics for 2,4-D degradation (Bandala et al., 2007b).

Table 1. Reaction rate values for the non photo-assisted Co/PMS and Fenton processes.

Co/PMS process		Fenton reaction	
Reaction conditions (mM)	k (min^{-1})	Reaction conditions (mM)	k (min^{-1})
$[Co^{2+}]$= 0.004; [PMS]= 1	0.003	$[Fe^{2+}]$= 0.004; $[H_2O_2]$= 1	-
$[Co^{2+}]$= 0.04;[PMS]= 1	0.012	$[Fe^{2+}]$= 0.04; $[H_2O_2]$= 1	-
$[Co^{2+}]$= 0.25; [PMS]= 1	0.060	$[Fe^{2+}]$= 0.1; $[H_2O_2]$= 1	0.003

As shown, by the increase in cobalt concentration the degradation efficiency and rated constants of 2,4-D increased. For example, for initial cobalt concentrations of 0.004, 0.04 and 0.25 mM, the ratios of rate constant were 1, 3.5 and 4.6 respectively. On the other hand, no 2,4-D degradation was observed for non photo-assisted (dark) process when low reagent conditions were tested. As show in table 1, only by using the highest reaction conditions we were able to determine low degradation reaction, as low as the lowest CO/PMS conditions (0.003 min^{-1}).

As described previously for solar driven AOPs, by the application of photo-assisted processes the overall efficiency of the pesticide degradation was improved by this mean. Table 2 show the results obtained by the use of photo-assisted AOPs, again considered as following a first order kinetics (Bandala et al., 2007b).

Table 2. Reaction rate values for the photo-assisted Co/PMS and Fenton processes.

Co/PMS/UV process Reaction conditions (mM)	k (min^{-1})	Photo-Fenton reaction Reaction conditions (mM)	k (min^{-1})
$[Co^{2+}]$= 0.0; [PMS]= 1	0.039	$[Fe^{2+}]$= 0.0; $[H_2O_2]$= 1	-
$[Co^{2+}]$= 0.004;[PMS]= 1	0.129	$[Fe^{2+}]$= 0.04; $[H_2O_2]$= 1	0.004
$[Co^{2+}]$= 0.25; [PMS]= 1	0.228	$[Fe^{2+}]$= 0.1; $[H_2O_2]$= 1	0.011

In order to discriminate the effect of temperature in the reaction rate for both photo-assisted processes, we have tested separately the influence of this parameter in the pesticide degradation. Results of the effect of temperature rising on the rate constant (*k*) for the tested procedures are depicted in table 3. Results in tables 2 and 3 show that the presence of solar radiation caused a significant increase in the reaction rate constant, much higher than the obtained only by the increase of temperature. For example, in the case of Co/PMS process, the increase in temperature to 40°C resulted in a *k* value of 0.0089 min^{-1}, over twice faster than the obtained in the dark Co/PMS system at 20°C. By the application of solar radiation, including the UV and IR radiation, *k* value increase up to 0.128 min^{-1} (35 times the value obtained for dark Co/PMS at 20°C).

Table 3. Reaction rate values, as function of different temperatures, for the photo-assisted Co/PMS and Fenton processes.

Co/PMS/UV process*		Fenton reaction**
Temperature (°C)	*k* (min^{-1})	*k* (min^{-1})
20	0.0036	0.0038
30	0.0045	0.011
40	0.0089	0.0604

* Experimental conditions tested were [cobalt]=0.004mM; [PMS]=1 mM. ** Experimental conditions tested were $[Fe^{2+}]$=0.1 mM; $[H_2O_2]$=1mM.

The effect of temperature on the Fenton reaction is also important as showed in table 3. The rise in the temperature caused the increase in the final 2,4-D degradation in a even more important way for the Fenton reaction than those showed for the Co/PMS process. Whereas for Co/PMS an increase of 20°C, from 20 to 40°C, produced an increase from 17 to 42% in pesticide degradation, when using Fenton reaction the same increase in temperature enhanced the final pesticide degradation from 17 to 95%. In the case of kinetic data, the same trend was determined, by increasing temperature from 20 to 40°C, the *k* value increased from 0.0038 to 0.06 min^{-1}, over one order of magnitude. The accomplished data, at different temperatures, for both Fenton and Fenton-like processes allowed us to determine the activation energy using the Arrhenius equation for both processes. The Co/PMS showed activation energy of 34.3 kJ/mol whereas the Fenton reaction's activation energy was estimated in 105.15 kJ/mol. With these results we are able to propose that Co/PMS process, with the lowest activation energy, is faster than the Fenton reaction as we have been able to observe from the different results showed in this work and that solar radiation is a real, interesting alternative to carry out the treatment of SESW wastewater.

Conclusions

Surfactant enhanced soil washing seems to be a real technological alternative for site restoration. Nevertheless few approaches are currently published about the efficient treatment of soil washing generated wastewaters. As showed in this work, complexity of the wastewater matrix could involve highly refractory and/or toxic components which avoid the direct use of biological processes for wastewater treatment.

AOPs appear as emergent technologies that could be useful for SESW wastewater treatment capable to perform interesting depletion in the concentration of toxic components which could lead to a easier biodegradability of the effluents. As it has been demonstrated here, high surfactant concentrations as well as total hydrocarbons and pesticides were successfully degraded using solar driven AOPs such Fenton-like processes.

Besides the importance of the application of solar driven AOPs to the treatment of SESW waste water, the possibility of use this kind of technologies not as the only wastewater treatment process but coupled to other conventional wastewater treatment process, i.e. biological processes, is a very interesting scientific and technological issue that must be accomplished before the application of SESW technology to site restoration at full scale.

References

[1] Anipsitakis G., Dionysiou D.D. 2003. Degradation of organic contaminants in water with sulfate radicals generated by the conjunction of peroxymonosulfate with cobalt. *Environmental Science and Technology* 67, 4790-4797.

[2] Anipsitakis G., Dionysiou D.D. 2004. Transition metal/UV-based advanced oxidation technologies for water decontamination. *Applied Catalysis B: Environmental* 54, 155-163.

[3] Bandala E.R., Aguilar F., Torres L.G.2009. Surfactant enhanced soil washing for the remediation of sites contaminated with pesticides. Land Contamination and Reclamation (In press)

[4] Bandala E.R., Martínez D., Martínez E., Dionysiou D.D. 2004. Degradation of microcystin-LR toxin by Fenton and Photo-Fenton process. Toxicon 43 (1), 829-832.

[5] Bandala E.R., Estrada C. 2007. Comparison of solar collection geometries for application to photocatalytic degradation of organic contaminants. *Solar Energy Engineering* 129, 22-26.

[6] Bandala E.R., Dominguez Z., Rivas F., Gelover S. 2007a. Degradation of atrazine using solar driven Fenton-like advanced oxidation processes. *Journal of Environmental Sciences and Health B: Pesticides* 42, 21-26.

[7] Bandala E.R., Peláez M.A., Dionysiou D.D., Gelover S., García A.J., Macías D. 2007b. Degradation of 2,4-dichlorophenoxyacetic acid (2,4-D) using cobalt-peroximonosulfate in Fenton-like process. *Journal of Photochemistry and Photobiology A: Chemistry* 186, 357-363.

[8] Bandala E.R., Pelaez M.A,, Salgado M.J., Torres L.G. 2008a. Degradation of sodium dodecyl sulphate in water using solar driven Fenton-like advanced oxidation processes. *Journal of Hazardous Materials* 151, 578-584.

[9] Bandala E.R., Velasco Y., Torres L.G. 2008b. Decontamination of soil washing wastewater using solar driven advanced oxidation processes. *Journal of Hazardous Materials160(2-3), 402-407.*

[10] Bandala E.R., Pelaez M.A., García-Lopez J., Salgado M.J., Moeller G. 2008c. Photocatalytic decolourisation of synthetic and real textile wastewater containing benzidine-based azo dyes. *Chemical Engineering and Processing* 47, 169-176.

[11] Chacon J.M., Leal M.T., Sánchez M., Bandala E.R. 2006. Solar photocatalytic degradation of azo-dyes by photo-fenton process. *Dyes and Pigments* 69, 144-150.

[12] Cheng H., Sabatini D.A. 2002. Simultaneous uptake of anionic surfactants and micellar-solubilized contaminants using anion exchange resins. *Water Research* 36(8), 2062-2076.

[13] Chu W., So W.S., 2001. Modelling the two stages of surfactant-aided soil washing. *Water Research* 35, 761-767.

[14] Chu W., Chan K.H. 2003. The mechanism of the surfactant-aided soil washing system for hydrophobic and partial hydrophobic organics. *Science of Total Environment* 307, 83-92.

[15] Gelover S., Leal T., Bandala E.R., Román A., Jiménez A., Estrada C. 2000. Catalytic photodegradation of alkyl surfactants. *Water Science and Technology* 42, 101-106.

[16] Iturbe R.,. Flores C, Chavez C., Ramirez A., Torres L.G. 2004a. In situ flushing of contaminated soils from a refinery : organic compounds and metal removals. *Remediation. Srping*: 141-152.

[17] Iturbe R., Flores C., Chavez C., Bautista G., Torres L.G.. 2004b. Remediation of contaminated soil using soil washing and biopile methodologies at field level. *Journal of Soils and Sediments*.

[18] Lambert T.W., Lane S. 2004. Lead, arsenic and polycyclic aromatic hydrocarbons in soil and house dust in the communities surrounding the Sydney, Nova Scotia, tar ponds. *Environmental Health Perspectives* 112(1), 35-41.

[19] Lee D.H., Chang H.W., Cody R.D. 2004. Synergism effect of mixed surfactant solutions in remediation of soil contaminated with PCE. *Environmental Geosciences* 8(3), 319-323.

[20] Lopez J., Iturbe R. ,Torres L.G.. 2004. Washing of soil contaminated with PAHs and heavy petroleum fractions using two anionic and one ionic surfactant: effect of salt addition. *Journal of Environmental Science and Health.* A39(9): 2239-2306.

[21] Mata J., Karns J., Torrents A. 2002. Influence of rhamnolipids and Triton X-100 on the desorption of pesticides from soils. *Environmental Science and Technology* 36, 4669-4675.

[22] Orozco S.L., Bandala E.R., Arancibia C.A., Serrano B., Suarez R., Hernandez I. 2008. Effect of iron salt on the color removal of water containing the azo-dye reactive blue 69 using photo-assisted Fe(II)/H_2O_2 and Fe(III)/H_2O_2 systems. *Journal of Photochemistry and Photobiology* 198, 144-149.

[23] Sabatini D.A., Harwell J.H., Hasegawa M., Knox R.C. 1998. Membrane processes and surfactant-enhanced subsurface remediation: a critical review. *Journal of Membrane Science* 15(1), 89-100.

[24] Santacruz G., Bandala E.R., Torres L.G. 2005. Chlorinated pesticides (2,4-D and DDT) biodegradation at high concentrations using immobilized Pseudomonas fluorescens. Journal of Environmental Science and Health B 40, 571-583.

[25] Sarria V., Kenfack S., Guillod O., Pulgarin C., 2003. An innovative coupled solar-biological system ar field pilot scale for the treatment of biorecalcitrant pollutants. *Journal of Photochemistry and Photobiology A: Chemistry* 159(1), 89-99.

[26] Scott M.J., Jones M.N. 2000. The biodegradation of surfactants in the environment. *Biochemical and Biophysical Acta* 1508 (1-2), 235-251.

[27] Torres L.G., Aguirre A.L., Verdejo A., Iturbe R.. 2005. Enhanced soil-washing treatment for soils which are highly contaminated with crude oil. In: Ecosystems and Sustainable Development V. *WIT Transactions on Ecology and the Environment.* 81: 541-550.

[28] Torres L.G., Orantes J.L., Iturbe R. 2003. Critial micellar concentrations for three surfactatnts and their diesel-removal efficiencies in petroleum-contaminated soils. *Environmental Geosciences* 10(1), 28-36.

[29] Torres L.G., Zavala P., Beltran B., Vaca M., Iturbe R. 2007a. Combination of natural gums and synthetic surfactants for washing of soil highly contaminated with crude. *Environmental Geosciences* 14(1), 1-9.

[30] Torres L.G., Climent M., Saquelares J., Bandala E.R., Urquiza G., Iturbe R. 2007b. Characterization and treatability of a contaminated soil from an oil exploration zone. *International Journal of Environmental Science and Technology* 4(3), 311-322.

[31] Torres L.G., Bandala E.R. 2008. Integrated surfactant enhanced soil washing and physical-chemical/biological treatment of produced wastewater. In: El Nemr A. (Ed.) New developments in environmental pollution and climate change. STM Books. Transworld Research Network (In press).

[32] Urum K., Pekdemir T., Ross D., Grigson S. 2005. Crude oil contaminated soil washing in air sparging assisted tank reactor using biosurfactants. *Chemosphere* 60, 334-343.

[33] Yuan T., Marshall W.D. 2007. Optimizing a washing procedure to mobilize polycyclic aromatic hydrocarbons (PAHs) from a field-contaminated soil. *Industrial Engineering and Chemistry Research* 46, 4626-4632.

[34] Zhang L., Somasundaran P., Osokov V., Chou C.C. 2001. Flotation of hydrophobic contaminants fro soil. *Colloids and Surfactants A:* 177, 235-246.

In: Remediation of Soils and Aquifers
Editors: Luis G. Torres and Erick R. Bandala
ISBN 978-1-60741-372-1

Chapter 3

ADVANCES ON PHYTOEXTRACTION

Margarita V. Beltrán, Mabel M. Vaca and Abelardo A. González.
Universidad Autónoma Metropolitana-Azcapotzalco

INTRODUCTION

Many years before the colonial times, the native Mexicans were knowledgeable about basic metallurgical techniques which were utilized to manufacture jewelry and ornaments. With the Spaniards arrival, exploitation of Mexico's mining potential acquired large scale proportions, situation that went on right through the twentieth century. Logically, mining exploration and exploitation kept on adapting to provide industry with new and better materials according to the advances of the steel industry processes and techniques. In some regions of the Mexican territory gold, silver and iron were extracted and processed using minerals which their mining ores were usually associated with other heavy metals and during the mineral enrichment process tons of tailings were produced. These residual materials were disposed inadequately due to a lack of knowledge of the environmental and toxicological implications that they would produce.

Heavy metals from the inadequate disposal of slags and tailings, in addition to the trace elements that are naturally found in high concentrations at mining sites, are susceptible to reach the man by drinking water, vegetables or animals in the food chain, promoted by processes such as solubility, volatility or lixiviation among other.

Records of the Federal Environmental Protection Bureau of Mexico (*Procuraduría Federal de Protección al Ambiente*, PROFEPA) reveal that arsenic (As), barium (Ba), cadmium (Cd), chromium (Cr), mercury (Hg), nickel (Ni), lead (Pb), selenium (Se) and zinc (Zn), are the most frequently heavy metals found in polluted soils (PROFEPA, 2000). The most probable pollution sources of these elements are: mining and foundry; agricultural materials; wastewaters; fossil fuel combustion; metallurgical industries; wasted metallic tools; manufacture, use and disposal of electronic hardware; the chemical and other manufacturing industries; industrial wastes; hunting; military and warfare materials (Alloway, 1995).

Soil pollution by heavy metals is important because contaminants may flow through the ground and pollute groundwater, they can also damage local vegetation and their toxicity may harm the human health. The number of polluted sites in urban has been increasing, and even in areas far from pollution sources it may be possible that children ingest a daily amount of heavy metals (Ljung, 2006). High concentrations of heavy metals have been found not only around mining and smelters, but also in agricultural soils, enriched with phosphorus fertilizers (Ericsson, 1990).

Recently 61 sites have been identified in Mexico as polluted with heavy metals; 23 of them are polluted with lead. It is said that this situation was a result of the years that passed without regulations for heavy metals pollution control, during decades of industrialization, residues were dumped in indiscriminate and uncontrolled ways because there was no infrastructure to properly handle this wastes (DOF, 2007).

SOIL RESTORATION TECHNIQUES

Implementation of efective techniques and procedures to treat polluted sites in order to reduce or eliminate contaminants concentration is the basis for pollution mitigation and restoration programs.

Restoration is defined as all activities undertaken to recover and re-establish conditions that allow evolution and continuity of natural processes. Specifically, in Mexico, soils' restoration is undestood as the set of actions undertaken to recover and re-establish its conditions, thus the contaminated area could be used again, without limitations, in any of the activities foreseen in the Urban Development Program or in the Environmental Territorial Management Plan* defined for that particular region (DOF, 2003).

Soil restoration has the objective to reduce the concentration of pollutants or to control them within a range that does not threat human health or would damage the environment, and it can be accomplished by (DOF, 2007):

- Permanent reduction of pollutants concentrations
- Reduction of bioaccesibility and/or solubility of pollutants
- Prevention of pollutants dispersion in the environment
- Setting up of institutional controls

Technologies applied to restore a polluted site, are based on:

- Physical-chemical treatments. They depend on physical or chemical procedures to remove soil pollutants, such as: precipitation, desorption, soil washing and ion exchange.
- Biological treatments. They are applied to remove, destroy or modify soil pollutants. It can be by biorestoration, composting, phytorestoration, or land farming among other techniques.

* The Environmental Territorial Management Plan is one of the environmental policy instruments of local governments to rule the use of land according to its potential and its characteristics. The Urban Development Program is another planning instrument that is applied in Mexico to regulate land uses, specifically designed for urban areas.

- Thermal treatments. They are applied to remove, stabilize or destroy soil pollutants, by vitrification, pyrolisis and/or thermal fixation.

The type of contaminants and their concentration, are the factors to be considered in order to select the proper technology to treat a specific hazardous waste. Treatment technologies are clasified according to the treatment mechanism applied as physical-chemical, biological or thermal technologies; regarding their location, they are classified as on site ("*in situ*") or out of the site (*"ex situ"*). Most of biological technologies are *in situ*, while the physical-chemical processes may be implemented *in situ* or *ex situ* (Tang, 2004).

Phytorestoration is defined as the use of vegetation to treat, *in situ*, soils, sediments and water that have been polluted with organic matter, nutrients and metals. It is suitable to treat shallow pollution (less than 3 m, depending on the depth of the roots) where it produces the best results. This method is applicable where other soil restoration techniques cannot be used for economic reasons, where pollution is at low level or where only a "superficial treatment" is required and there is no time restriction to achieve the expected results (from 6 months to 3 years to achive the complete cleaning process). It can be used together with other techniques for better results, and the final vegetation cover resulting from phytorestoration technique can also avoid site erosion (Schnoor, 1997).

Table 1. Phytorestoration mechanisms and techniques (adapted from Singh *et al.*, 2003).

Phytorestoration technique	Mechanism	Applicable to	Media
Rizodegradation	Contaminant removal at plant's root	Metals, explosive residuals and some organic compounds	Surface water and water pumped through the roots
Phytotransformation	Nutrients and organic compounds assimilation	BTEX, chlorinated solvents and short chain aliphatic compounds	Soil, surface and groundwater
Rhizosphere biodegradation	Biodegradation by microorganisms in rhizosphere	Nitroaromatic compunds such as TNT, halogenated hydrocarbons such as chlorinated solvents and pesticides, and anilines	Soil and groundwater surrounding the rhizosphere
Phytoextraction	Direct pollutant absorption in plant tissue followed by plant's removal	Heavy metals and radionuclides	Soil
Phytostabilization	Root's exudes cause precipitation, then biomass inmovilizes them and reduces their bioavailability	Heavy metals and radionuclides	Groundwater, soil and tailings
Phytovolatilization	Plants evaporate metal ions, organic compounds or their metabollic products	Heavy metals and organic compounds	Soil, groundwater

Phytoremediation has been promoted as a promising non-destructive and *in situ* clean-up process (Crowley *et al.*, 1997, Lutts *et al.*, 2004). The term phytorestoration incorporates a variety of techniques which include different mechanisms to remove toxic materials from the environment as shown in Table 1.

Phytoextraction

In phytoextraction, heavy metals in contaminated soils are translocated to aboveground parts of selected species, thus reducing its concentration in soils to acceptable levels.

Some plants absorb larger quantities of metals in comparison with others. For phytoextraction purposes a specific type of plant is selected regarding the contaminants to be removed and the site characteristics. The selected plants are sow and after they are developed, they are cut and incinerated, or they are left to transform into vegetal fertilizer to recycle metals. This procedure can be repeated as many times as required in order to reduce the concentration of pollutants to acceptable levels. If plants are incinerated, residues must sent to a controlled disposal site, but the amount of the ashes to be disposed of is only about 10% of the total volume of the residue that would be eliminated if the contaminated soil was excavated for *ex situ* treatment.

The use of hyper-accumulating plants, together with adequate agronomic practices in order to maximize their biomass production has been proven attractive. The hyper-accumulating plant concentrates at least 0.1% of the element in its aerial parts (though this percentage varies depending on the element being accumulated) (Reeves and Baker, 2000).

Plants use different mechanisms to tolerate heavy metals present in soils. One of such mechanisms is hyper-accumulation, which implies transportation of metals from soil to new tissues, in concentrations up to 10 mg Hg/kg, 100 mg Cd/kg, 1000 mg/kg of Co, Cu, Cr, Pb, or greater than 10,000 mg/kg of Zn and Ni (Chaney *et al.,* 1995, Schnoor, 1997). These plants can be harvested, air-dried and incinerated and the heavy metals recovered in the ashes can be sustainability handled (Chaney *et al.,* 1999).

The metals best suited for phytoextraction are Ni, Zn, and Cu, because they are preferred by more than 400 plant species that are known as high metal absorbants (EPA, 1996).

An important issue in the phytoextraction process is related to the disposal method of metals. They can be recovered from plant tissue, or they have to be disposed as hazardous waste. Some design considerations involve the "accumulation factor" (defined as the ratio of metal content in plant tissue with respect to metal in soil) and the productivity of plants (biomass dry-weight that can be harvested each cycle). In order to have an efficient treatment, aplant with intensive growth (dry weight biomas produced should be greater than 3 tons/ha/year) that can be easily harvested and that accumulates large quantities of metal in the harvesting portion of the plant (i. e. more than 1,000 mg/kg, dry weight) is required. To achieve a clean up process between 3 to 5 years, plants must accumulate metals levels 10 times greater than current the amount in the contaminated soil. For instance, if the soil shows a metal concentration of 500 mg/kg, the corresponding concentration in the plant must be about 5000 mg of metal/kg biomass dry weight (Schnoor, 1997).

To improve the phytoextraction process, the bioavalability of pollutants to the roots can be increased using acidifying agents, fertilizers or chelates (Prasad and Freitas 2002).

As a general rule, metals such as Cd, Ni, Zn, As, Se and Cu are more bioavailable for plants than Co, Mn and Fe. On the other hand, Pb, Cr and U are the metals less bioavailable for plants, but they can be removed by rizofiltration with plants with high mass density of roots. If EDTA is added to the soil (0.5-10 μg EDTA/kg of soil), it may improve Pb bioavailability, however, this will also increase Pb solubility and, therefore, the probability that Pb migrates to groundwater (Wu *et al.*, 1995).

Plants do have efficient mechanisms to form metallic complexes in the cytosol utilizing organic compounds produced by the plants, and then the vacuoles store the metals. Recent research has uncovered some of the mechanisms of tolerance and accumulation of metals, particularly with the Zn hyper-accumulating species *Thlaspi caerulescens* (Becerril *et al.*, 2007).

Formation of these metallic complexes is due to the presence of chelating compounds, such as metalotioneins and phytochelatines, which play a role in the detoxification of plants. Synthesis of these compounds is induced by the root cells exposed to these types of pollutants. A cysteine enriched polypeptide that is present in plants creates metallic bonds with the thiol group of proteins in order to achieve vegetal tissue detoxification. The metalotioneins are sulfur rich proteins formed with 60 to 80 amino acids.

Phytochelatines belong to the γ-glutamylcysteine oligopeptides family with glycine or other carbonyl amino acids that are synthesized from glutathione and its derivates by the phytochelatine synthetase in the presence of heavy metals. The proposed mechanism of formation of metallic complexes by phytochelatines to be kept inside the vacuole is presented in Figure 1 (Peuke and Rennenberg, 2005).

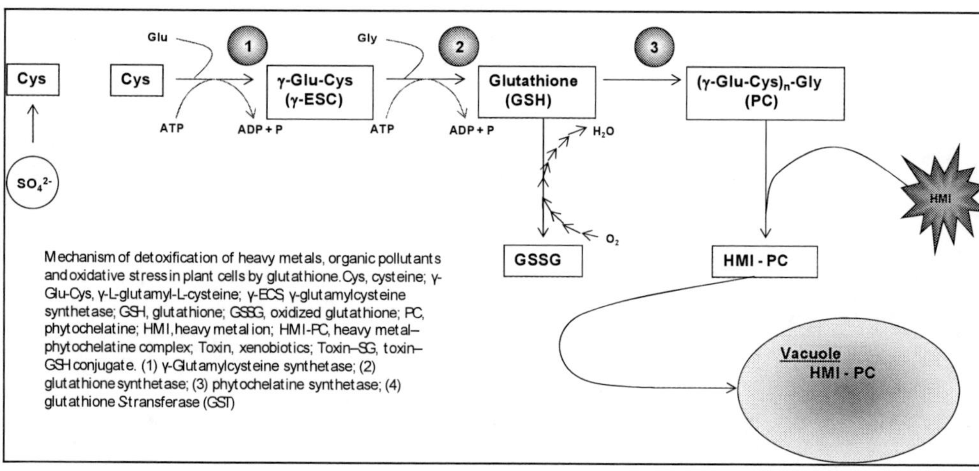

Figure 1. Diagram of detoxification of heavy metals, organic pollutants and oxidative stress in plant cells (Adapted from Peuke and Rennenberg, 2005).

A PHYTOEXTRACTION CASE STUDY: MIXTURE OF GRASS UTILIZED TO RESTORE AN INDUSTRIAL-URBAN SOIL

Due the high toxicity and persistence of heavy metals in the environment, soil restoration is one of the biggest challenges that environmental engineering has to face (Wu *et al.*, 2004). A phytoextraction case study with experimental rizotrons placed in a greenhouse where an industrial soil polluted with heavy metals was treated utilizing a mixture of different grass species, is presented in this section. The objectives of the project were to evaluate the heavy metals assimilation capacity and their concentration in grass' biomass, and to asses the reduction of heavy metals concentrations in the polluted soil as a restoration process.

An industrial-urban soil mainly polluted with Ni and Cu was used for these experiments. The pollution of this soil is a result of fifty years of uncontrolled infiltration process that has allowed the pollutants to reach 15 m depth. Soil tested came from the surface layer of the site; it was characterized and, per kg of soil dry weight, it had approximately 2,500 mg of Ni, 4,000 mg of Cu, 350 mg of Pb, 300 mg Zn and 8 mg of Cd.

Copper is an essential mineral for plants growth and development, and it is present in a range between 23 to 140 ppm, with an average of 88 ppm; whereas in industrial affected areas Cu may be in the range between 1,500 and 20,000 ppm; these are toxic to vegetation. Even though there is no Mexican regulation regarding Cu contents in soil, the observed values for the test soil were higher than the limits that most vegetation tolerates naturally (Kabata-Pendias y Pendias, 1992). The Canadian Council of Ministers of the Environment (CCME) recommends a maximum total Cu concentration of 63 mg/kg of soil for agricultural, recreational parks or residential land use; whereas for commercial or industrial land use the maximum recommended concentration is 91 mg/kg (CCME, 2006).

Nickel is naturally found on earth's crust combined with other elements; it is a micronutrient needed for plant development, therefore its presence in soil is necessary. In non polluted soils Ni is found normally within a range of 4 to 80 ppm; whereas in sites surrounding extraction areas Ni concentration can reach up to 9,000 ppm. The CCME recommends a Ni concentration of 50 mg/kg of soil for agricultural, parks, residential commercial or industrial use (CCME, 2006). The NOM-147-SEMARNAT/SSA1-2004 is the Mexican rule that describes the criteria for restoration of soils that have been polluted with arsenic, barium, beryllium, cadmium, chromium VI, mercury, nickel, silver, lead, selenium, thallium and/or vanadium; and, for soils that will be used for agricultural or domestic applications, Ni concentration must be less than 1,600 mg/kg; whereas for land that will be use for industrial purposes, Ni is limited to 20,000 mg/kg.

Cadmium is also naturally found on earth's crust as part of rocks, coal minerals and it is even found in manure. The average Cd content varies from 1 to 450 ppm (dry weight) in soils of different countries. In Canada, it has been reported the existence of sites where the Ni pollution reached up to 26,000 ppm (dry weight) and in New Zealand it has been found locations with Cd pollution level from 1,700 to 5,000 ppm (Kabata-Pendias and Pendias, 1992).

Lead is a natural constituent of soil and dust. Natural Pb level is in the range of 10 to 50 mg/kg (dry weight) but human activities may increase these level 10 to 200 times (Albert, 1997). It has been found that Pb content can reach up to 2,000 mg/kg within a distance of 1 to 25 m from main roads, and these contents decline exponentially with distance; in areas surrounding foundries Pb has been measured as high as 60,000 mg/kg dry weight (Kabata-Pendias and Pendias, 1992).

Zinc is a common element in earth's crust, it can be found in the air, soil and water, and it is present in food. In small quantities, it is an essential element for animal life. The natural Zn content in soil's surface of different European, Asian and American countries range from 17 to 125 ppm (dry weight). In England it has been reported 66,400 ppm of Zn in polluted soils and levels of 53,000 ppm have been reported in the United States, though Zn pollution level normally is between 1,000 to 10,000 ppm dry weight (Kabata-Pendias and Pendias, 1992).

Methodology

The soil sample for the experimentation was taken at surface level and came from a polluted urban-industrial site in the metropolitan area of Mexico City. Soil's characterization was carried out at the beginning of the experiment, before the grass was sown, and it comprised the following parameters: pH, porosity, organic matter, cationic exchange capacity (CEC), texture, humidity and water content at field capacity, nutrients content (N, P, K) and heavy metals.

Soil's physical-chemical characterization indicated that Cu, Ni, Zn, Pb and Cd concentrations exceeded the normal content in non-polluted soil.

A mixture of grass species (*Festuca arundinacea, Festuca rubra genuina, Lollium perenne, Agrostis tenuis and Festuca rubra fallax)* was sown because it had been observed that this approach let a better utilization of the individual advantages of each grass variety and the mixture provides a higher resistance to many environmental factors and plagues (González, 2007). The experiment was conducted sowing the grass mixture in rizotrons filled with approximately 5 kg of polluted soil that had been previously prepared with urea (Figure 2).

Figure 2. Experimental rizotrons.

Monthly soil and plant's tissue samples were taken in triplicates from reactors with two different experimental conditions (blank and polluted soil) during a four-month period (2 experimental conditions x 4 months x 3 repetitions). It was considered that testing procedure would be destructed; therefore reactor's content was discarded after sampling and analysis.

Contaminated soils were then characterized, and final pH (EPA, 1995), porosity, texture, nutrients, organic matter, cationic exchange capacity and Cd, Cu, Ni, Pb and Zn concentrations were determined. Heavy metals in soils were analyzed after acid digestion as indicated in EPA method 3050A (EPA, 1992), using atomic absorption spectrometry with a Varian Spectra AA-200 apparatus. Detection limits were 0.02, 0.2, 0.02, 0.02 and 0.2 mg/L for Cd, Pb, Ni, Cu and Zn, respectively, using the corresponding EPA methods (EPA, 1986).

With regard to biomass determination, the amount of stem and leaf material from each plant was weighed. Then a certain amount of the material was measured and digested in 5mL 50% (v/v) trace metal grade nitric acid in a boiling water bath. Extra nitric acid was added as needed to maintain the total volume at 5 mL. For roots and leaves, a statistical analysis was performed to determine which data points should be removed due to instrumental error.

Results and Discussion

Greenhouse environmental conditions were monitored during the 120 days of the experimental period. The information is presented in Table 2.

Table 2. Grenhouse environmental conditions.

Environmental condition		Record
Temperature, °C	Minimum	20°C ± 2
	Maximum	40°C ± 7
	Average	29°C
Relative humidity	Minimum	33 ± 3%
	Maximum	88 ± 10%
	Average	53 %

Table 3. Soil characterization.

Parameter	Conditioned polluted soil	Blank soil
pH	6.57 ± 0.11	8.05 ± 0.15
Texture	20% clay, 30% silt, 50% sand	87.55% silt, 12.5 % clay
Organic matter, %	2.17 ± 0.35	3.63 ± 0.43
Field capacity water content, %	58.3 ± 4.1	64.7 ± 4.2
Nitrogen	.097 ± 0.001 % 970 ± 10 ppm	0.23 ± 0.02 % 2300 ± 200 ppm
Phosphorus, ppm	58.0 ± 3.2	89.5 ± 2.9
Potassium, ppm	84.0 ± 7.0	39.01 ± 0.24
CEC, meq Na/100 g soil	14.34 ± 0.63	13.51 ± 1,5
Porosity, %	63.3 ± 2.20	60.78 ± 0.52

Physical-chemical analysis results of the experimental soil are presented in Table 3. Polluted soil was mildly acid whereas the blank was alkaline (DOF, 2000); the texture of polluted soil was silty, and the blank presented a silty clay texture; regarding organic matter content, it was determined that polluted soil was medium rich, while blank soil was rich in organic matter (Reyes, 1996).

Both soil samples presented a typical cationic exchange capacity (10<CEC<25 meq Na/100 g of soil) (Reyes, 1996). In relation to soil's nutrients, the available phosphorous content determined by BRAY-I method for both soils was high (>30 ppm); nitrogen content was medium in the polluted soil, and it was rich in the blank soil; potassium content was medium in the polluted soil, and it was low for the blank soil (Reyes, 1996).

Heavy Metals

Heavy metals results of the soil's experiments are presented in Table 4 where they are compared with the total reference concentration (TRC) for a soil that is used for industrial purposes as defined in the Mexican regulation NOM-147-SEMARNAT/SSA1-2004 (DOF, 2007), and with the reported average for heavy metals contents in hystosol. Heavy metals concentration of the polluted soil was higher than the range indicated for organic soils (hystosol).

Table 4. Heavy metals concentration in experimental soils.

Heavy Metal	Conditioned polluted soil	Blank soil	TRC	Hystosol average [a]
Cadmium, mg/kg	7.9 ± 1.6	ND	450	0.19-2.2
Copper, mg/kg	4,105.5 ± 100	24.0 ± 2.8	NR	1-113
Nickel, mg/kg	2,586.7 ± 83.8	4.35 ± 0.05	20,000	0.2-119
Lead, mg/kg	357.0 ± 37.9	52.85 ± 5.8	800	1.5-176
Zinc, mg/kg	327.5 ± 18.9	92.9 ± 11.0	NR	5-250

NR: Not reported
ND: Undetected
CRT: Total reference concentration as defined in the NOM-147-SEMARNAT/SSA1-2004
[a] Reference range for organic soils. Hystosol average (Kabata-Pendias and Pendias, 1992)

For polluted sites, Cu contents have been reported as high as 3,700 ppm (mg Cu/kg soil dry weight); 2,000 ppm Pb have been found in urban sites and up to 60,000 ppm Pb near foundries; Ni content has been reported in the range of 1,700 to 5,000 ppm; and Zn pollution in soils varies from 1,000 to 10,000 ppm (Kabata-Pendias, 2001). Comparing these results with Table 4, it can be observed that copper content is higher than reports, while Ni, Pb and Zn are comparable with the information reported elsewhere for polluted sites.

Vegetation Growth

During experimentation period (120 days) grass growth and biomass production were measured monthly. Lengths of roots and the aerial part of the plants were measured as shown in Figure 3. For both soil experimental conditions the roots increased their length during the whole experimentation period; while the aerial part of the plant showed a constant length increase during the first 90 days, and then, during the last 30 days, its growth was asymptotical. The maximum length was observed at 90 days, 22 cm in contaminated soil and 50 cm in the blank; whereas root's length observed in the blank was 52 cm, and 19 cm the in polluted substrate. Dry weight biomass production of the grass mixture in the contaminated soils showed 45% inhibition as compared to the blank.

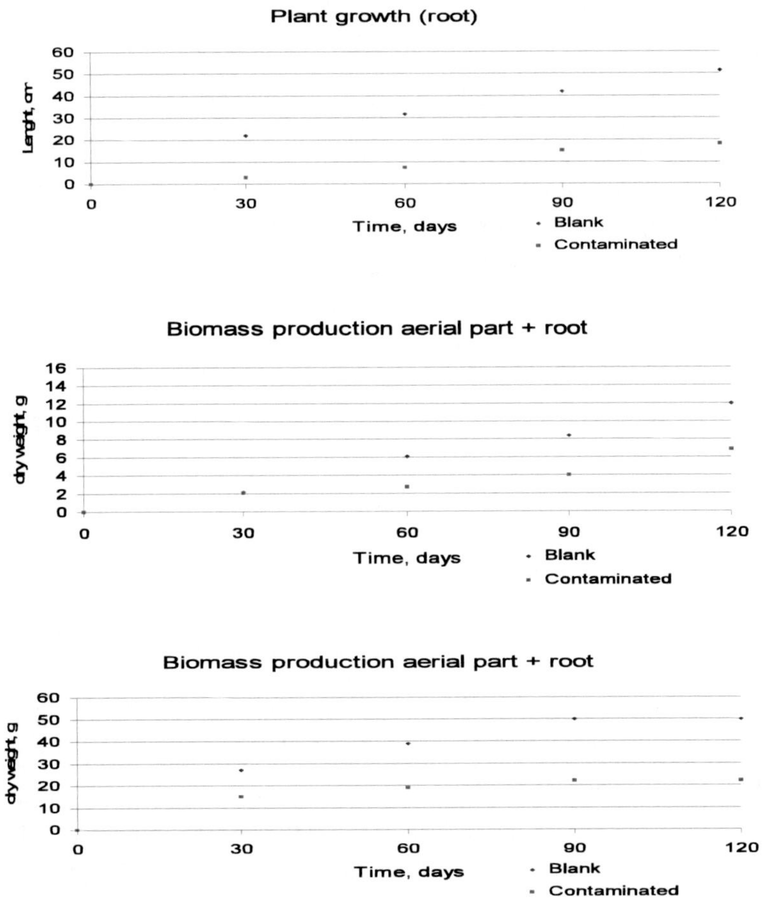

Figure 3. Monthly monitoring of grass mixture.

Phytoextraction Process Evaluation

Heavy metals extraction after 120 days of experimentation period is shown in Table 5.

Heavy metals removal during experimentation period is presented graphically in Figure 4; the removal order was as follows: Cd>Zn>Cu>Pb>Ni; although the higher removal efficiency was with Cd, the concentration of this metal was the lowest, and there is a relationship between metal removal efficiency and the initial and final concentration in soil.

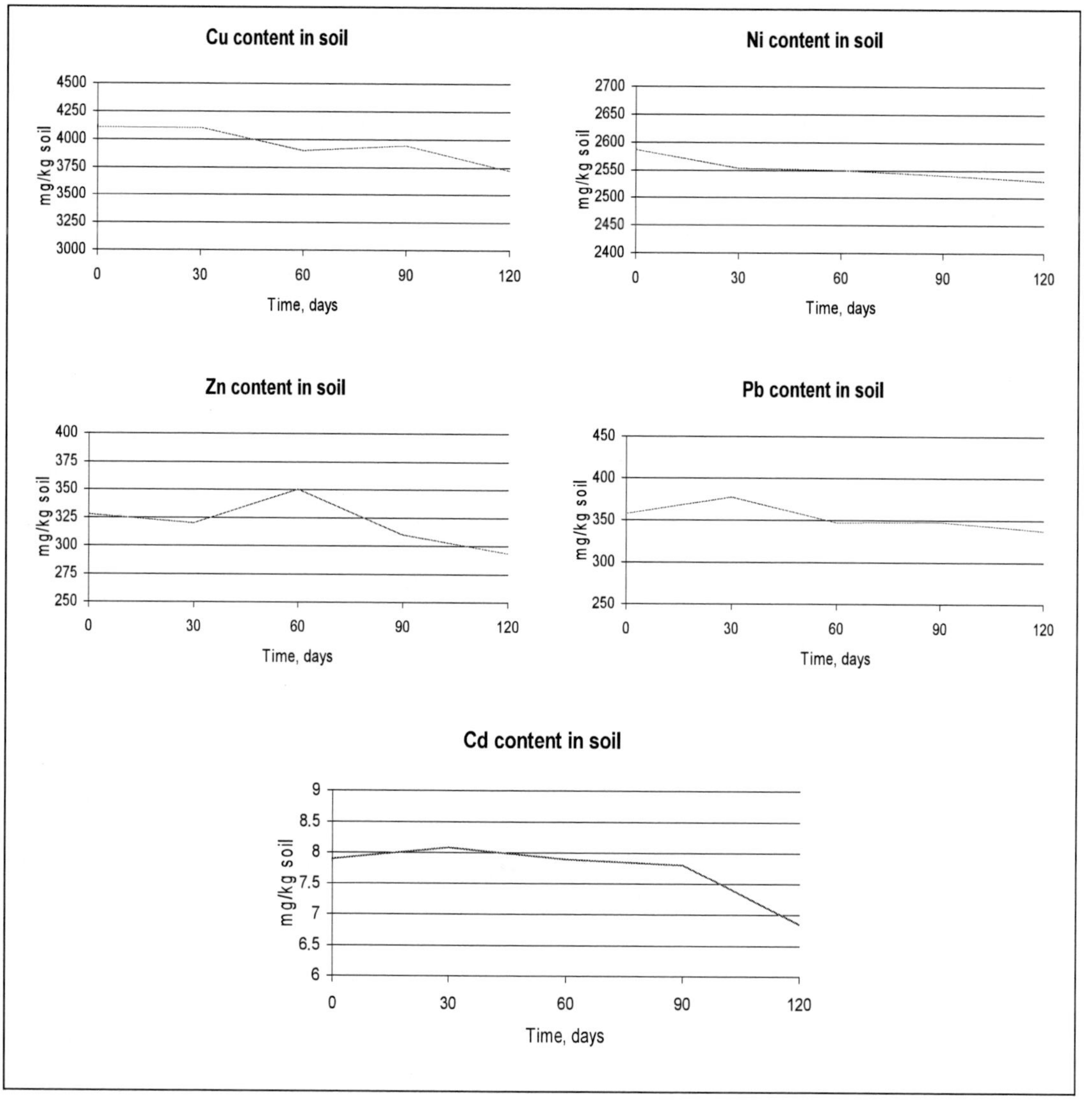

Figure 4. Heavy metals concentration during experimentation period.

Table 5. Heavy metals' removal with the grass mixture after 120 days.

Heavy Metal	Initial concentration, mg/kg dry weight	Final concentration, mg/kg dry weight	Removal efficiency, %
Cu	4 105.5 ± 100	3 721.8 ± 120	9.35
Ni	2 586.7 ± 83.8	2 530.9 ± 52.9	2.16
Zn	327.5 ± 18.8	293.5 ± 10	10.36
Pb	357 ± 8	337.8 ± 17	5.37
Cd	7.9 ± 1.6	6.8 ± 0.7	13.24

Heavy metals concentrations in plant tissue during experimentation period are presented in Figure 5. Regarding Cu concentration, in this experiment it increased during the first 60 days in both samples (root and aerial part of the plant), then it showed an asymptotical behavior. The average Cu content in grass that was grown in non-polluted sites was 10.5 mg/kg of biomass in dry weight, whereas in contaminated regions coffee stems with Cu content as high as 1,122.5 mg/kg of biomass in dry weight have been found (Kabata-Pendias, 2001). *Aeollanthus bifomifolius* has been identified as Cu hyper-accumulation plant because Cu contents up to 13,700 ppm have been reported (Chaney et al., 1999). For the grass mixture used in the experiment, Cu content was 1,750 mg /kg of total biomass in dry weight, taking into consideration the proportions found in root as well as in the aerial part of the plant.

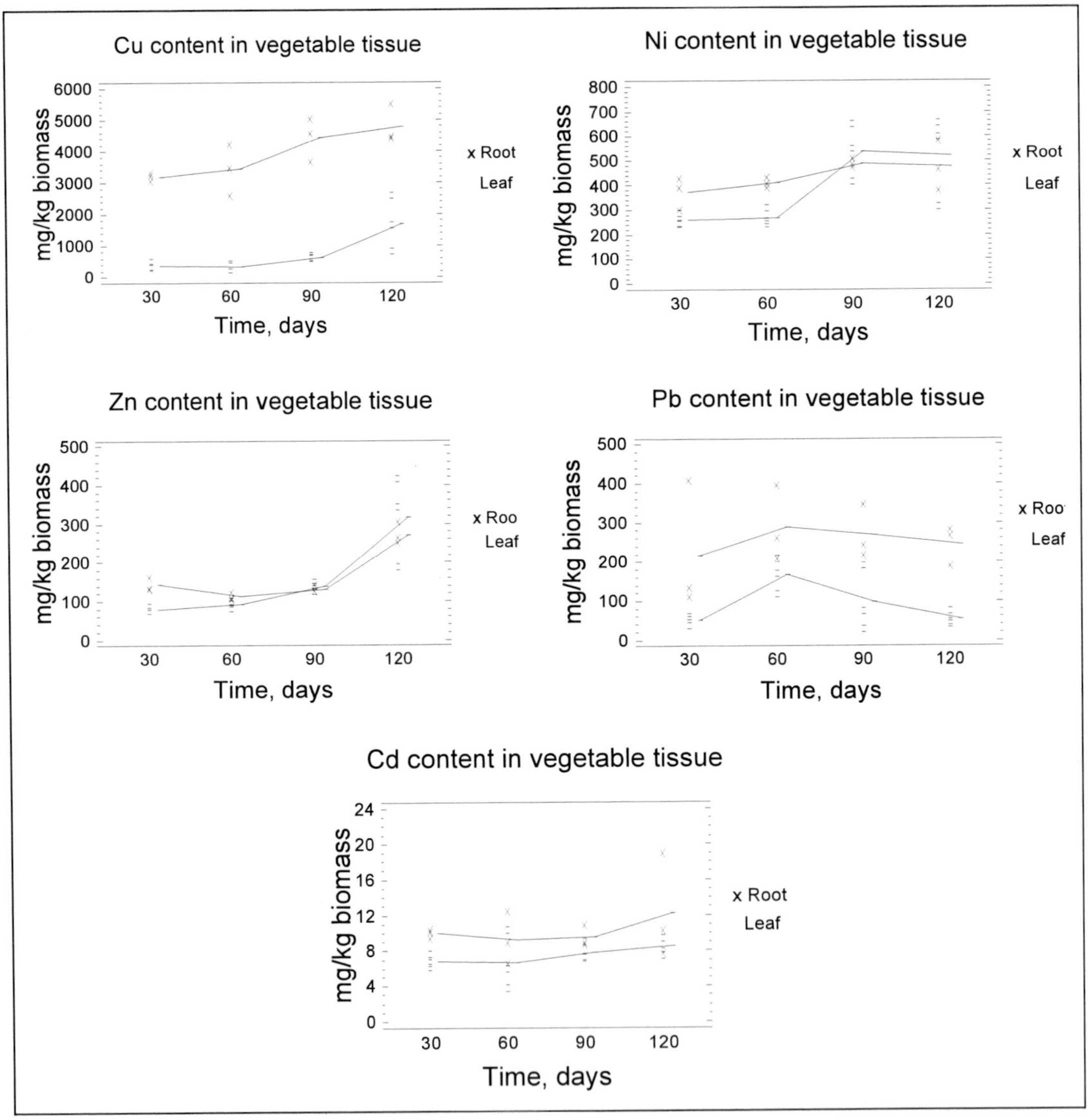

Figure 5. Heavy metals concentrations in plant tissue during experimentation period.

In the case of Ni, the accumulation was higher in the root than in the aerial part of plants. Average Ni content in grass from non-polluted sites has been reported within the range of 0.1 to 1.7 ppm, but for grass from polluted sites it has been found from 1,700 to 32,000 ppm

(Kabata-Pendias, 2001), and for the grass mixture used in the experiment, Ni content was up to 425 mg Ni/kg of total biomass in dry weight.

From the beginning of the experimentation Zn concentration was higher in the root than in the aerial part of the plant, and its content increased during experimentation time. Literature reports for rice that grows in contaminated sites present Zn concentrations as high as 4,510 ppm (Kabata-Pendias, 2001), and *Thlaspi caerulenscens* can accumulate up to 51,600 mg Zn/kg of biomass in dry weight (Cunningham and Ow, 1996). The results of this experiment showed a concentration of 230 mg Zn/kg of total biomass dry weight.

Cd in the experiment was concentrated mainly in the roots and there was very little translocation to plant's aerial part. By the time the experimentation concluded, Cd accumulation in the grass mixture biomass was 11.3 ppm dry weight. Normal Cd contents in grass from non-polluted sites has been reported between 0.07 and 0.27 ppm, and *Thlaspi caerulenscens* has been recognized as Cd hyper-accumulating plant with a metal content of up to 1,800 ppm dry weight of biomass (Cunningham and Ow, 1996).

In this experiment, the higher proportion of Pb was accumulated in the root and very little movement to the aerial part of the plant was observed. Lead content in the grass mixture of the experiment reached a concentration of 125 mg Pb/kg of total biomass dry weight. As a reference, normal Pb content in grass that grows in non-polluted sites is 2.1 ppm and it has been reported that *Thlaspi caerulenscens* can grow in contaminated sites and concentrate up to 2,714 ppm of Pb in its biomass (Kabata-Pendias, 2001).

Heavy metals accumulation in plants presented the following order: Cu>Ni>Zn>Pb>Cd. The metals accumulation level is directly related with soil's concentration, thus the higher the metal concentration in the soil the higher the accumulation in plants. This was true with the exception of Zn that had a lower concentration than Pb in the soil, but in the plants it presented higher contents. Heavy metals accumulation in both biomass sections that were analyzed for after the end of the experimental period is presented in Table 6.

Table 6. Heavy metals content in biomass after 120 days of experimentation period.

Heavy Metal	Concentration in leaf, mg/kg dry weight	Concentration in root, mg/kg dry weight
Cu	796.91 ± 100.8	4958.33 ± 190
Ni	313.92 ± 42.8	522.52 ± 82.6
Zn	187.75 ± 25	321.26 ± 24.7
Pb	60.43 ± 15	227.27 ± 26
Cd	8.14 ± 0.9	7.53 ± 1.5

CONCLUSIONS

The comparison of the amount of each metal accumulated in roots and leaves is presented in Figure 6. There is a clear indication of the amount of metal that is present in each part of the plant, and even though there were higher metals concentrations in the roots, the absolute amount of metal accumulated in the plant was higher in the aerial part due the amount of biomass produced in this section. It is worth mentioning that in the case of Cu, after three months the amount of metal accumulated was almost the same in the root than in the aerial part of the plant.

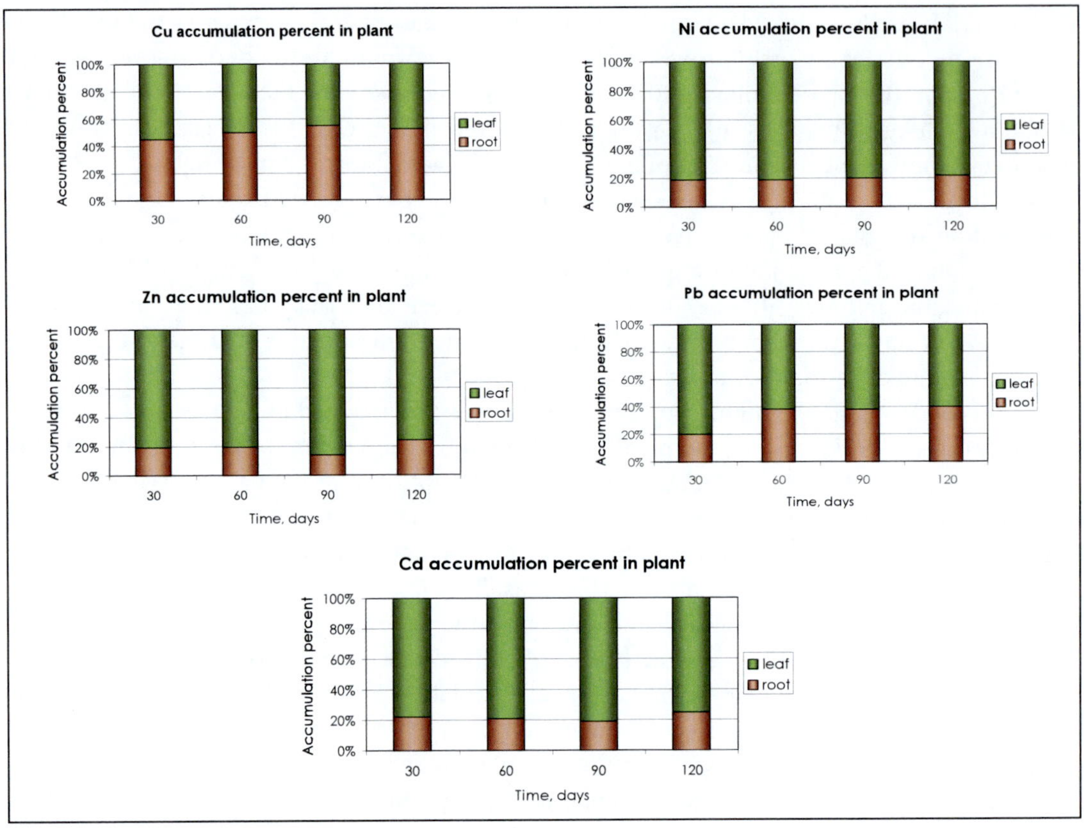

Figure 6. Heavy metals accumulation percentage in leaves and roots.

This study evaluated heavy metals phytoextraction process in an industrial polluted site. Copper, nickel, zinc, lead and cadmium were the metals present in the contaminated soil. A grass mixture was used to evaluate phytoextraction process.

Metals uptake and plant assimilation were directly proportional to their concentration in soil. The metals concentrations in plants, expressed as mg of metal/kg of plant tissue in dry weight, were 1,725 of Cu, 425 of Ni, 230 of Zn, 125 of Pb and 11 of Cd.

The removal efficiencies after 120 days were 9.35%, 2.16%, 10.36%, 5.37% and 13.24% for Cu, Ni, Zn, Pb and Cd respectively.

All heavy metals analyzed in this study accumulated in high concentrations in roots; however the mass of metal was higher in the aerial part of the plant due to the high biomass production rate. The plants seeded in polluted soil showed a growth inhibition effect of approximately 45%, as compared to the same plants that were planted in the reference soil.

Grass mixture used in this experiment can be classified as Cu hyper-accumulating because it could accumulate in its biomass more than 1,000 ppm of metal; for the other metals analyzed in this study this criterion was not complied. The grass mixture used was able to assimilate, tolerate and accumulate high heavy metals concentrations in the soil, and it was also able to produce acceptable quantities of biomass.

CASE STUDIES ON PHYTOEXTRACTION

The resistance of plants to the presence of high concentrations of metals responds to two different strategies: evasion and tolerance; the first one prevents an excessive metal uptake, while the second one is the capacity to withstand an excessive metal accumulation in their tissues (Cervantes and Moreno, 1999). The search for plant species which are able to accumulate high heavy metals concentrations implicate studies that must be focused in the evaluation of concentrations to produce a phytotoxic effect (plant's growth reduction), their tolerance and heavy metals bioaccumulation capacity, and identification of factors that may affect the assimilation of metals by the plant.

In order to evaluate plant assimilation capacity sometimes the metal concentration in the plant is compared to known values of hyper-accumulating plant species; this ratio is called the relative assimilation index (RAI). In the search of plant species for restoration processes, the use of the RAI has the advantage that, being a relative number which is related with the accumulation of a specific contaminant in certain plant species, any plant with a RAI closer to one may be considered suitable for such pollutant. RAI also represents the ratio of the concentration that is present in plant's biomass in relation with the metal concentration in the indicator plant (Arroyo, 2003).

Table 7 shows results from several phytoextraction studies that had been carried out in Mexico City by the Metropolitan Autonomous University (Universidad Autónoma Metropolitana: UAM, by its initials in Spanish). Food plant species were evaluated by Beltrán and coworkers (2006), and they found that the romaine lettuce and the sunflower transported more Cd to stems and leaves, while the *crispa* lettuce and the zucchini accumulated it mainly at the roots. The studied species had an acceptable performance in the phytoextraction process under the tested conditions because of their tolerance and hyper-accumulation of these metals (Cd and Zn) in the vegetable tissue; however, the edible characteristic of the plants offered a tangible risk, and strict management measures were recommended related to their environmentally-safe final disposal and the prevention of their consumption.

Díaz (2001), Arroyo (2003) and Cortés et al. (2007, 2008), studied different tolerant and heavy metal accumulation plant species, and in some cases they related their findings with respect to lettuce as metal hyper-accumulation plant species.

Cárdenas (2002), Arciniega (2002) and Vaca et al. (2006) have evaluated heavy metals assimilation potential of plants that have been sown in polluted soils conditioned with biosolids as soil's enhancer. Biosolids usually contain heavy metals within the permissible limits stated in Mexican regulation NOM-004-SEMARNAT-2002 (DOF, 2003).

Ayala (2007) developed a diagnosis of the quality of a urban-industrial soil polluted with heavy metals at different soil layers, and a study of the metal concentration in the plants surrounding the study site. The presence of As, Cu and Ni in concentrations above the permissible limits established in the Mexican regulation for restored sites: NOM-147-SEMARNAT/SSA1-2004 (DOF, 2007) was confirmed.

Even though phytoextraction uses plant species to retain inorganic contaminants through the roots and to accumulate them in stems and leaves, the study of this extraction process can be directed as treatment technology: (1) to reduce pollutants concentration in soils; (2) as a sequential post-treatment after the soil has been physicochemically treated in order to

Table 7. Selected studies on phytoextraction

Pollution source	Plant species	Study type	Results	Reference
Artificially contaminated soil with $Cd(NO_3)_2 \cdot 4H_2O$ and $ZnCl_2$ to obtain two different experimental conditions: the first one with 150 mg Cd/kg of soil dry weight; and the second one with 150 mg Cd/kg of soil dry weight and 350 mg Zn/kg of soil dry weight	Squash (*Cucurbita moschata*) Lettuce (*Lactuca sativa* var. *longifolia* and *Crispa* var.) Sunflower (*Helianthus annus*)	To evaluate Cd and Zn assimilation capacity and plant's uptake under controlled greenhouse conditions. To produce information for health's risk evaluation	Concentrations in squash up to 357 ppm Cd (mg Cd/kg dry tissue) in roots, 212 ppm in stems, 204 ppm in leaves, 18 ppm in flowers and 12 ppm in fruits were observed. The most efficient species in the phytoextraction process was the lettuce, competing with squash (in the soil with Cd) and sunflower and squash (in the soil with Cd-Zn). Due metals' tolerance and bioaccumulation these plant species represent a health risk according with the Commission of the European Communities (2001) and the Codex Alimentarius Commission (2004)	Beltrán *et al.*, 2006
Artificially contaminated soil with $Cd(NO_3)_2 \cdot 4H_2O$ and $Pb(NO_3)_2$ to obtain a soil with 150 mg Cd/kg of soil dry weight and 2,000 mg Pb/kg of soil dry weight	Lettuce (*Lactuca sativa* var. *longifolia* and *Crispa* var.) *Lupinus* var. *Russell* *Zinnia elegans*	RAI determination of plant species under study. Test soil presented high organic matter (18%). Initial and final heavy metal concentrations were studied. Effects in soil's physicochemical properties were studied after 130 days of experimentation.	Lead content in lettuce, *Lupinus* and *Zinnia* was 850, 167 y 60 mg de Pb/kg dry tissue respectively; whereas Cd content was, in the same order, 728, 224 y 62 mg de Cd/kg dry tissue. Observed Pb RAI was 1.0 for lettuce, 0.07 for *Lupinus* and 0.196 for *Zinnia*; Cd RAI observed was 1.0 for lettuce, 0.085 for *Lupinus* and 0.31 for *Zinnia*.	Arroyo, 2003

Pollution source	Plant species	Study type	Results	Reference
Silty-clay soil from the outskirts of a mining zone mixed with slags in Actopan, Hidalgo. Soil had 2,970 mg As/kg dry soil; 147 mg Cd/kg of dry soil; and 323 mg Pb/kg drysoil.	Turnip (*Brassica napus*) Radish (*Raphanus sativus)*	Bioconcentration and As, Cd and Pb selectivity determination in plant species after a 4 weeks experimentation period	Survival rate for radish was 100% and for turnip was 17%. When radish was used, 50% removal efficiency for the three contaminants was achieved in rihzosphere soil. It was determined that radish presented a 100 mg Cd/kg of dry new tissue (leaves and fruits), in roots, it was measured about 1000 ppm for As and also for Pb.	Díaz, 2001
Polluted industrial soil with the following initial concentrations (dry weight): 2,100 mg Cu/kg soil; 1,800 mg Ni/kg soil; and 700 mg As/kg soil.	Lettuce (*Lactuca sativa* var. *longifolia* and *Crispa* var.) Sunflower (*Cosmo bipinnatus*) Grass (*Poa pratense*)	Heavy metals' tolerance and extraction capacity of plant species. Feasibility for phytoextraction process. Metals under study were Cu, Ni and As. Experiment's duration: 100 days.	In lettuce it was found Cu concentrations up to 1,500 ppm (mg Cu/kg dry biomass), 680 ppm Ni and 1,100 ppm As; in sunflower the metals' concentrations were 900 ppm Cu and 770 ppm Ni; whereas for grass it was observed a concentration up to 2,700 ppm Cu and 580 ppm Ni; no As was detected in sunflower and grass. Metal´s extraction was measured as the relationship between extracted metal to initial soil concentration, and itas was 2.8% for Ni, 12% for As and between 2.5 and 10.6% for Cu.	Cortés *et al*., 2007

Table 7. Continued

Pollution source	Plant species	Study type	Results	Reference
Biosolids conditioned with zeolite (clinoptilolite) and phosphoric rock and their applicability in agricultural uses. Biosolid content respect to soil vary from 5 to 100	Lettuce (*Lactuca sativa* var. *Intybacea* Hort)	Biomass generation and metal concentration in plant tissue after 25 days experiment.	Biosolid's metals content (mg/kg dry biosolids) was 3.2 Cd, 48 Cr, 236 Cu, 18 Pb and 33 Zn. Plant sown in conditioned soil produced as much biomass as the one in nutrient enriched soil. The concentration of metal reported in plant tissue was normal.	Cárdenas, 2002. Vaca *et al.*,2006.
Soil conditioned with 3% of biosolids. Metal's concentration, as mg/kg dry biosolids, were 0.3 As, 1.7 Cd, 445 Cr, 86 Pb, 445 Cu and 987 Zn.	Squash (*Cucurbita moschata*)	Biomass generation and metal concentration in plant tissue after 120 days experiment.	30 to 60% in excess biomass was produced in biosolids conditioned soil when compared with reference value of nutrient enriched soil. Metals concentrations, as mg/kg dry biomass, were 0.2 As, 23 Cr, 22 Cu, 52 Pb, 78 Zn and Cd was not detected.	Arciniega, 2002.
Urban-industrial polluted soil. Initial metal concentrations, as mg/kg soil dry weight, were 1,800 Ni, 700 As and 2,100 Cu.	Aloe (*Aloe vera barbadensis miller*)	Heavy metals extraction and assimilation. Metals under study were Cu, Ni and As. Experiment's duration: 100 days.	Metal's presence in soil inhibited up to 75% the vegetative development when compared with the reference test. Metal concentrations in plant's tissue, as mg/kg dry biomass, were 750 NI, 720 As and 1,300 Cu. Metal accumulation proportion was higher in roots, more than 50% for Ni, and more than 70% for both As and Cu.	Cortés *et al.*, 2008.

Pollution source	Plant species	Study type	Results	Reference
Urban-industrial polluted site. Metal concentrations observed, as mg/kg soil dry weight, were: Cu from 17 to 29,222 ppm, Ni from 45 to 2,010 ppm, and As from 3.8 to 4,580, at different soil's layers in which depth vary from 0 to 1.5 m, 1.5 to 3.2 m, and 3.2 to 5.0 m. The highest concentrations were observed in 3.2 to 5.0 m soil layer.	Ash treee, (*Fraxinus uhdei*) Elm tree, (*Ulmus minor*) Coral tree, (*Erytrina americana Mill*) Sweetgum, (*Liquidambar styraciflua*) Casuarina, (*Casuarina equisetifolia*) Camellia, (*Camellia japonica*) Eucalipts, (*Eucalyptus camaldulensis*) Cypress, (*Cupressus sempervirens L.*) Willow, (*Salix bonplandiana H. B. K.*)	Pollutant's content in flora of the study area.	Heavy metals were determined in all tree leafs of the study area. Metal´s concentrations, as mg/kg dry biomass, were up to 60 ppm Ni, 472 ppm Cu and 85.4 ppm As. These metal´s contents exceed the normal average concentration reported for these plant species.	Ayala, 2007.

achieve, at low cost, the permissible limits; (3) to stabilize soils and to prevent metals' lixiviation to lower layers or groundwater; and (4) to form a vegetation cover to avoid surface erosion and pollutant transport by air.

Besides the use of phytoextraction as restoration technique, studies may also focus in the evaluation of the potential health risk of plant species that grow in polluted sites, which are part of the human food chain; and to diagnose pollution impact in plant species that are found in contaminated sites.

REFERENCES

[1] Albert, L.A., (1997). Introducción a la toxicología ambiental, Centro Panamericano de Ecología Humana y Salud, México.

[2] Alloway, B. J. (1995). Heavy Metals in Soils. 2nd ed. Blackie Academic and Professional, Glasgow and London.

[3] Arciniega, C. R., (2004), Factibilidad del uso de biosólidos acondicionados con zeolita para su aplicación en la agricultura. Proyecto Terminal en Ingeniería Ambiental, Universidad Autónoma Metropolitana Azcapotzalco, México.

[4] Arroyo, Z., I., (2003), Índice de Asimilación Relativa (IAR) de metales pesados en especies vegetales, Proyecto Terminal en Ingeniería Ambiental, Universidad Autónoma Metropolitana Azcapotzalco, México.

[5] Ayala, Q. I., (2007), Diagnóstico de calidad de un suelo industrial contaminados con metales pesados y su efecto en la vegetación, Proyecto Terminal en Ingeniería Ambiental, Universidad Autónoma Metropolitana Azcapotzalco, México.

[6] Becerril J. M., Barruita O., García Plazaola J. I., Hernández A., Olano J. M., Garbisu C., (2007), Especies nativas de suelos contaminados por metales: aspectos ecofísicos y su uso en fitorremediación, Ecosistemas, Asociación Española de Ecología Terrestre.

[7] Beltrán V. M., Vaca M. M., López C. R., Alcalá H. J. M. (2006). Fitoextraction of heavy metals and transport and distribution in plant tissue The Second International Meeting on Environmental Biotechnology and Engineering (2IMEBE) 26-29 September, Mexico City, Mexico.

[8] Cárdenas, V. J. I., (2002), Manejo ambiental de biosólidos para uso agrícola. Tesis de Maestría en Ciencias e Ingeniería ambientales, Universidad Autónoma Metrpolitana Azcapotzalco, México.

[9] Cervantes, C., Moreno, S. R., (1999). Contaminación ambiental por metales pesados. Impacto en los seres vivos, AGT Editor S.A. México. D. F.

[10] CCME (Canadian Council of Ministers of the environment) (2006). Canadian Soil Quality Guidelines for the Protection of Environmental and Human Health, Julio.

[11] Chaney, R.L., Brown, S.L., Angle, J. S., Baker, A. J. M. (1995). Zinc and cadmium uptake by hiperaccumulator Thlaspi caeurescens and metal tolerant Silene vulgaris grown on sludge–amended soils, *Environ. Sci. Technol.* 29 (6): 1581-1585.

[12] Chaney, R.L., Brown, S.L., Stuczynski, T. I., Daniels, W. L., Henry Ch. L., Li, Y. M., Siebielec, G., Malik, M., Angle, J. S., Ryan, J. A., Compton, H. (1999). In-situ remediation and phytoextraction of metals from hazardous contaminated soils, US-

EPA´s Conference on Innovative Clean-Up Approaches: Investments in Technology Development, Results & Outlook for the Future, Nov. 2-4, Bloomingdale, IL.

[13] Codex Alimentarius Commission (2004) Report of the 36th Session of the Codex Alimentarius Committee on food Additives and Contaminants, Rotterdam, The Netherlands, 22–26 March.

[14] Commission European Communities (2001), Commission of the European Communities-Commission Regulation (EC) No 466/2001 of 8 March 2001 setting maximum levels for certain contaminants in foodstuffs, *Official J. Eur. Communities* L77, 01–13.

[15] Cortés, J.E.V., Beltrán, V. M., González, E. B., Vaca M. M., Torres, B.L.G., (2007), Phytoextraction of Ni, As and Cu in polluted soil of urban/industrial sites. The Second Internacional Workshop on Biotechnology, CEVIDE, Pachuca, Hidalgo, México.

[16] Cortés, J.E.V., Beltrán, V. M., Vaca M. M., Torres, B.L.G., (2008), Asimilación de metales pesados en especies vegetales que crecen en un suelo contaminados. XVI Congreso Nacional de Ingeniería Sanitaria y Ciencias Ambientales. D.F., México.

[17] Crowley, D. E., Cunningham, S. D., Shaan, J. R., Anderson, T. A. (1997). *Phytoremediation of contaminated water and soil.* ACS Books, Washington, D.C.

[18] Cunningham, S.D., Ow, D.W., (1996). Promises and prospect of phytoremediation, *Plant. Physiol.* 110:715-719, E.U.A.

[19] Díaz, D.M., F., J., (2001), Fitoextracción de metales pesados en un suelo urbano aplicando rábanos y nabo, Proyecto Terminal en Ingeniería Ambiental, Universidad Autónoma Metropolitana Azcapotzalco, México.

[20] DOF, Diario Oficial de la Federación, (2000). Proyecto de Norma Oficial Mexicana PROY-NOM-021-RECNAT-2000que establece las especificaciones de fertilidad, salinidad y clasificación de suelos: estudios, muestreo y análisis, México.

[21] DOF, Diario Oficial de la Federación, (2003). Norma Oficial Mexicana NOM-004-SEMARNAT-2002, Protección ambiental.- Lodos y biosólidos.-Especificaciones y límites máximos permisibles de contaminantes para su aprovechamiento y disposición final. México.

[22] D.O.F. Diario Oficial de la Federación (2003), Norma Oficial Mexicana NOM-138-SEMARNEY/SS-2003, Límites máximos permisibles de hidrocarburos en suelos y las especificaciones para su caracterización y remediación.

[23] DOF, Diario Oficial de la Federación, (2007). Norma Oficial Mexicana NOM-147-SEMARNAT/SSA1-2004, que establece los criterios para determinar las concentraciones de remediación de suelos contaminados por arsénico, bario, berilio, cadmio, cromo hexavalente, mercurio, níquel, plata, plomo, selenio, talio y vanadio, México.

[24] EPA (1986) Cd, K, Zn, Hg, Pb, Fe, Cu, Mg, As, Ni, Cr (Atomic Absorption, Direct Aspiration), 7130, 7610, 7950, 7471, 7420, 7380, 7210, 7461, 7061, 7520, 7190 *Methods*. September.

[25] EPA (1992*). Acid Digestion of Sediments, Sludges, and Solis*, 3050a Method, July.

[26] EPA (1995). *Soil and Waste pH. 9045C Method.* June.

[27] EPA (1996) Guía del Ciudadano: Medidas fitocorrectivas. Organismo para la Protección del Medio Ambiente (E.U.). Desechos Sólidos y Respuesta en Situaciones de Emergencia (5102 G). EPA 542-F-96-025. Septiembre de 1996.

[28] Ericsson, J. E. (1990). A field study on factors influencing Cd levels in soils and grain of oats and winter wheat. *Water Air Soil Pollut.* 53 (1): 69-81.

[29] González, E., B., (2007). Fitoextracción en un suelo contaminado con metales pesados mediante pastos. Proyecto Terminal en Ingeniería Ambiental. Universidad Autónoma Metropolitana. Unidad Azcapotzalco. México.

[30] Kabata-Pendias, A., (2001). *Trace elements in Soils and Plants, 3a edición*, CRC Press.

[31] Kabata-Pendias, A., Pendias, H. (1992).*Trace Elements in Soils and Plants*, CRC Press.

[32] Ljung, K., Selinus, O., Otabbong, E., Berglund, M., (2006). Metal and arsenic distribution in soil particle sizes relevant to soil ingestion by children, *Applied Geochemistry,* 21 (9).

[33] Lutts,S.,Lefevre, I., Delperee, C., Kivits, S.,Dechamps, C., Robledo, A., Correal, E. (2004). Heavy metal accumulation by the halophyte species mediterranean saltbush, *J. Environ. Qual,* 33 (4): 1271-1279.

[34] Peuke A. D., Rennenberg H., (2005), Phytoremediation, EMBO reports, *European Molecular Biology Organization.*

[35] Prasad, M.N.V., Freitas, H.M.O., (2003). Metal hyperaccumulation in plants-Biodiversity prospecting for phytoremediation technology, *Electronic Journal of Biotechnology,* Vol. 6, No. 3, Chile.

[36] PROFEPA, (2000) Procuraduría Federal de Protección al Ambiente, Disposiciones y Procedimientos para la Caracterización y Restauración de Suelos Contaminados: lista de criterios interinos para inorgánicos tóxicos (Anexo III), México.

[37] Reeves, R. D.; Baker, A. J. M. (2000). Metal-accumulating plants. In Phytoremediation of toxic metals: using plants to clean up the environment; Raskin, I., Ensley, B.D., eds.; John Wiley & Sons: New York. 193–229

[38] Reyes J. I. (1996) Fundamentos teórico-prácticos de temas selectos de la ciencia del suelo. Parte I, Universidad Autónoma Metropolitana Unidad Iztapalapa, Libros de texto, Manuales de Prácticas y Antologías, México.

[39] Schnoor, J. L., (1997) Phytoremediation, Department of Civil and Environmental Engineering, University of Iowa.

[40] Singh O.V., Labana S., Pandey G., Budhiraja R., (2003), "Phytorremediation: an overview of metallic ion descontamination from soil", *Appl Microbiol Biotech*, 61: 405-.412.

[41] Tang, W. Z., (2004), Physicochemical treatment of hazardous wastes, 1a Edición, Lewis Publishers, EUA, 23-66.

[42] Vaca, M. M., Beltrán, V. M., Vázquez, M. A., Callejas L. R., Hachec, L. R., (2006). Fertilización dosificada con biosólidos acondicionados. Revista Internacional AIDIS, V.1, 1. ISSN 0718-378X .

[43] Wu, J., Hsu, F. C., Cunningham, S. D., (1995) Chelade- Assisted Pb Phytoextraction: Pb Availability, Uptake, and Traslocation Constrains. *Environ. Sci. Technol.,* 33 (11), 1898-1904.

In: Remediation of Soils and Aquifers
Editors: Luis G. Torres and Erick R. Bandala ISBN 968-1-60741-372-1

Chapter 4

USE OF GIS AS A TOOL TO DEFINE SOME HYDROGEOGICAL CHARACTERISTICS OF THE SANTA ALEJANDRINA WETLAND IN VERACRUZ, MEXICO

Alejandrina Castro, Diana Santa Arango and Rosario Iturbe
Instituto de Ingeniería-UNAM

INTRODUCTION

Mexico is a country with a positive water balance regarding aquifers; however, its water availability is unequally distributed. The North, with an arid to semidesertic climate, and the largest cities of the country, as well as a high concentration of industrial and agricultural activities, has less than one third of the water resources of the country (Marín, 2004). Information on the underground water resource of the country is scarce, of the 654 aquifers identified administratively (DOF, 2001), less than half of them have been studied regarding their water availability. Of the studied aquifers, 194 are overexploited and 17 are affected by marine water intrusion (CONAGUA, 2008). Hydrogeological studies in the country are scarce and isolated, hence, many of the decisions on the management of water resources lack technical support. These studies are important, because the collected hydrogeological parameters generate a conceptual model that describes the hydrogeological aspects of the site, this information, aside from helping in the management of the resource, allows selecting, if needed, the necessary remediation technique(s) for the corresponding aquifer.

Since 2003, diverse projects have been implemented by the Hydrogeology Management Agency of the Mexican Geological Service, related to the exploitation, supply, use, and sustainable management of the water resources, aimed at providing support to the hydric resource operating and regulating agencies (Salas, 2007). Wetlands, important constituents of the hydric resources, have also been scarcely studied in the country. Wetlands are highly dynamic sites on their own, subjected to a wide range of natural factors that determine their modification along time. Their physical attributes, mainly hydrographic, topographic, and edaphic, are constantly modified by endogenous processes such as sedimentation, and by exogenous phenomena, such as flooding (Lindig-Cisneros and Zedler, 2005). To assess and understand the behavior of a wetland it is important to obtain the largest amount of data on how ground water occurs, depths of the phreatic level, thickness of the materials and their lithology, as well as on water chemistry. A large part of these data were obtained from the Santa Alejandrina Swamp, the study area object of this work; however, we were unable to gather information on the topography of the site at a scale that would allow knowing the hydraulic loads of the site, hence it was necessary to use a Geographical Information System (GIS) to generate the digital model of the terrain (DMT) at a scale that would allow us to observe the piezometry of the site. To elaborate this model, we obtained information with differential GPS technology on the topographical elevations of five piezometer nidi and one piezometer. The objective of this study is to show the usefulness of the geographical information system (GIS) tools to generate data to describe the conceptual hydrogeological frame.

GIS (Theoretical Frame)

The Geographical Information Systems are used in multiple applications, to solve very diverse problems. There is a large amount of definitions, so that it could be stated, following Barredo's definition (1996), that the GIS technology has evolved since its conception in the same manner as its conceptualization.

The reported definitions on GIS are similar but diverse and establish that an information system is designed to work with spatially or geographical referred data. A GIS is then a mixture of human resources, information and digital geographical data, including data processing with the adequate hardware and software, that is used to respond to diverse requests on geographical trends (Vega, 2001).

GIS resulted from the evolution of Computer, Informatics, and Geography sciences. Georeferenced geographical data constitute the basis of the whole information system; without them neither the hardware nor the software would make any sense. Georeferencing means to assign some type of coordinates linked with the terrain to the objects of interest; these can be natural ones, engineering works, the vertices of a parcel, among others.

Nowadays, a large variety and amount of programs to work with Geographical Information Systems can be found in the software market according to the specific needs and objectives required by their users. Some of these systems are *MapInfo* and *ArcGIS*, although open-source software is also available.

GIS can be traced back to the fifties and, in Mexico, the interest in developing them arose in 1972, when the Public Education Ministry started to consolidate an Educational Planning System, which included spatial dimensioning of geographical type information. In this system, it was possible to retrieve information on whether the municipalities had enough teachers for their student population (Vega, 2001). Afterwards, the development of GIS was much slower; however, in the last years due to the international technological advance on this subject, great efforts have been made to trigger the GIS technology in diverse areas, such as Cadastral Modernization Programs, Urban and Regional Planning, Demographical and Statistical Analyses, Topography, Geodesics, Geology, Planning and Inventory of Natural Resources, Election Systems, Environmental Impact Studies, among the most important ones (Vega, 2001). The enterprise known as PEMEX (Mexican Oil Co.) counts upon the

Corporate Information System (SICORI®, for its initials in Spanish), since 1991, which provides support to the company for the planning and decision taking regarding the physical safety of the oil facilities (INEGI, 2000). The GIS used by the Comisión Federal de Electricidad (Mexican Agency in charge of providing electricity) helps in planning and provides support for the development of specific projects (Vega, 2001). The National Institute of Statistics, Geography and Informatics (INEGI, for its initials in Spanish) is continuously expanding the National System of Geographical Information, its objective, among others, is to produce cartographies. The National Water Commission (CONAGUA, for its initials in Spanish) uses GIS as support tool for the spatial analysis of the information regarding water. The National Council of Population is developing the Georeferenced Data Analysis Unit to analyze socio-demographic phenomena. The Federal Electoral Institute uses GIS as a support tool for the electoral geography and for operational logistics. Some state governments are developing GIS to be used mainly for regional planning and cadastral information. In the educational realm, the number of universities developing GIS for research purposes is continuously increasing, leading the way the Universidad Nacional Autónoma de México (Marín, 2008).

Among the many applications of GIS in Mexico, we can mention its use in hydrogeology projects (Salas, 2007). Determining the conceptual hydrogeological model is of great help in those sites needing remediation, as GIS can indicate the preferred migration patterns of contaminants and their depth.

As will be seen ahead, the study area is located in a wetland, which are sites undergoing continuous changes (Lyon and McCarthy, 1995) due to anthropogenic and natural processes (Moya et al, 2005). In these cases, the GIS are important tools that allow updating continuously the information that might change the initially proposed conceptual model, due to its capacity to gather, manipulate, and analyze the information.

GENERAL FEATURES OF THE SITE

The Santa Alejandrina swamp is located in the southeast of the state of Veracruz, in the municipality of Minantitlan, approximately 20 km southwest of the Coatzacoalcos Port (Figure 1), between coordinates 94°32'30" and 94°28'30" longitude and 17°57'30" and 18°00'00" latitude, covering an approximate surface of 23 km^2. The study zone is surrounded south and east by the Coatzacoalcos river, the main fluvial way of the region; to the north it is surrounded by flooding plains close to the coastal line and to the west by the city of Minatitlán. The area is within a zone of wetlands located in the left margin of the Coatzacoalcos river, where several PEMEX facilities, constituting an oil complex, are located. The zone pertains to the aquifer 3012, named Costera de Coatzacoalcos, state of Veracruz (CNA, 2002), which is a water body belonging to the Santa Alejandrina swamp.

The study area is located within the physiographic province *Llanura Costera del Golfo Sur*, which is a coastal plain of strong alluvial input from the rivers flowing through it to mouth in the south of the Gulf of Mexico. Two physiographic units are identified in this plain, one represented by high zones and the other by flat zones; the first occupies the central portion covering the higher portions of hilly terrain with elevations reaching 150 m asl, the second is drained by the last stretches of the Coatzacoalcos and Tonalá rivers, before they mouth in the Gulf of Mexico. Swampy areas corresponding to the lower zones and, consequently, floodable are found in this coastal plain (CNA, 2002). This flat relief characteristic of the area, together with the swampy conditions, leads to minimal movement of the surface water.

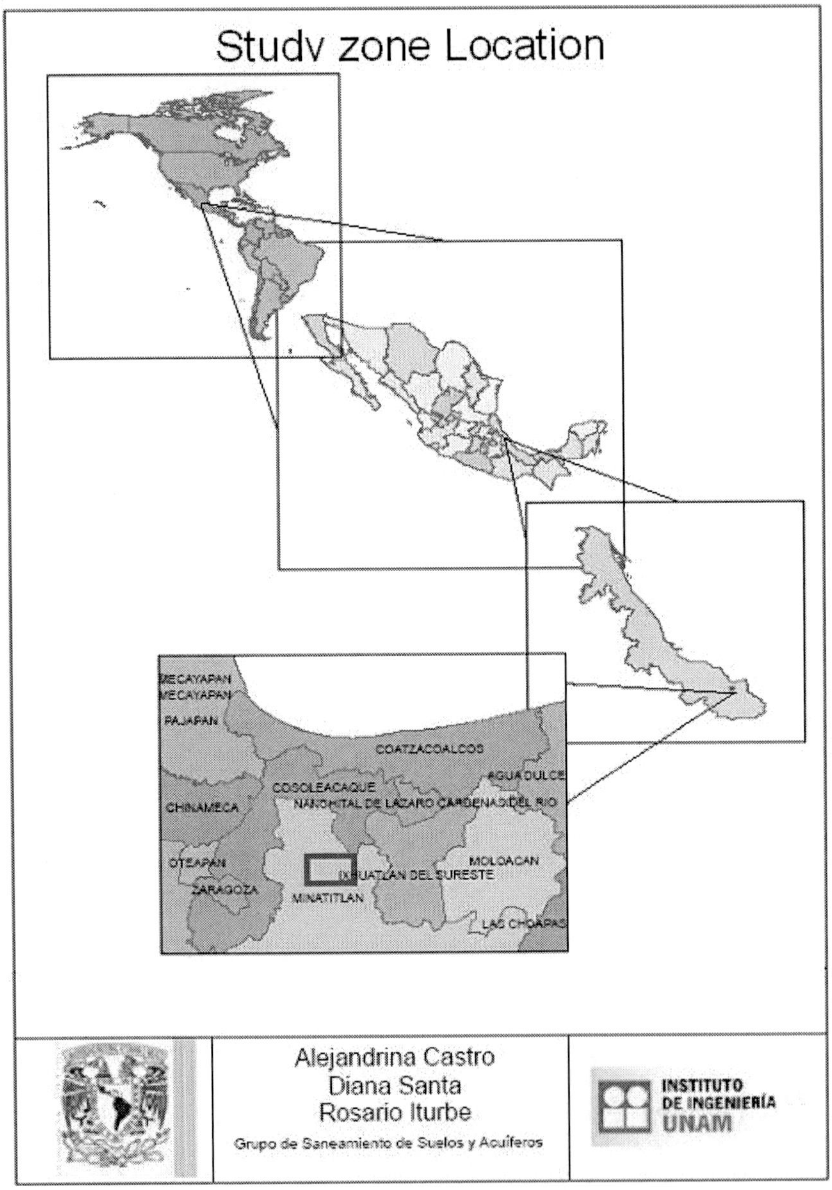

Figure 1. Santa Alejandrina swamp location

The climate is hot, humid and subhumid, with a mean annual temperature of 25 °C and a mean annual rainfall of 2041 mm. The swamp is part of the Hydrological Region No 29, Coatzacoalcos. The main rivers are the Coatzacoalcos and Tonalá (INEGI, 2007).

Regarding its geology, only the Quaternary is represented in the zone, constituted by flooding plain deposits composed of fine sands, limes, and non-consolidated clays (CNA, 2002).

In general, most of the hydrogeological studies performed in the aquifer 3012 have been circumscribed to the coastal plain of the Coatzacoalcos River. The main focus of these studies was to localize zones suitable for well bore drillings to exploit the underground water resources and to quantify preliminarily their volume. The main objective of the studies performed exclusively on the Santa Alejandrina swamp has been to quantify its current contamination, aside from treatability studies to determine which remediation techniques are the most adequate (Yañez-Trujillo *et al.*, 2003); because the Santa Alejandrina swamp presents a high degree of contamination by hydrocarbons because it has been used as disposal site of oily muds and worn-out catalizers generated from the nearby refinery, as well as of residual waters contaminated with hydrocarbons, which used to be discharged in the swamp towards the San Francisco stream, located to the northeast of the study zone. However, up to now, neither the degree of deterioration nor the nature and distribution of contaminants are fully known, it is only known that some areas have been subjected to treatment, which does not imply a complete remediation of the whole affected area.

Material and Methods

The GIS was generated based on data compiled from previous works and those collected in the field. Information was stored and its analysis and interpretation has allowed obtaining other type of data leading to the digital model of the terrain.

Compilation of Information

The National Institute of Statistics, Georaphy and Informatics (INEGI) has relevant information on the geology, hydrology, physiography, topography, and communication routes of the state of Veracruz. This information was used to elaborate the base plan of the study area. On its part, the Water and Sanitizing Municipal Commission of Minatitlán (CMASM, for its initials in Spanish) provided geophysical information and of the water quality of the city of Minatitlán. Geological information of the zone was also obtained from the Mexican Geological Service (SGM, for its initials in Spanish). Information on phreatic levels recorded in other years and availability of water from the regional aquifer was obtained from the National Water Commission (CONAGUA). The available information provided a preliminary approach to the study area, mainly regarding the hydrology of the site.

Field Work

Because the study area has no water-exploitation well, nine piezometers and five piezometer nidi were constructed, the latter at three different depths, 1.50 m, 1 m, and 0.5 m. Phreatic level measurements were performed in each piezometer, by means of an electrical

Solinst probe. Since the topography of the site is practically flat, leveling had to be performed by means of a differential GPS (Trimble, model 4700) and a mobile double frequency receptor (Trimble, model 4800),. For the post-processing of data, the Trimble Geomatics Office software was used. To achieve leveling, a base station was defined within the area and identified as "BASM"; leveling was only performed in the piezometer nidi. To optimize the quality control of the data obtained with the differential GPS, the base station was linked to a fixed station of the Active National Geodesic Network (RGNA, for its initials in Spanish) with an acquisition time of 9 hours and 27 minutes. This was the closest station to the study area, located in the state of Oaxaca, southwest of the study area (Figure 2). Post-processing generated the base lines; configuration of the system of coordinates is depicted in Table 1.

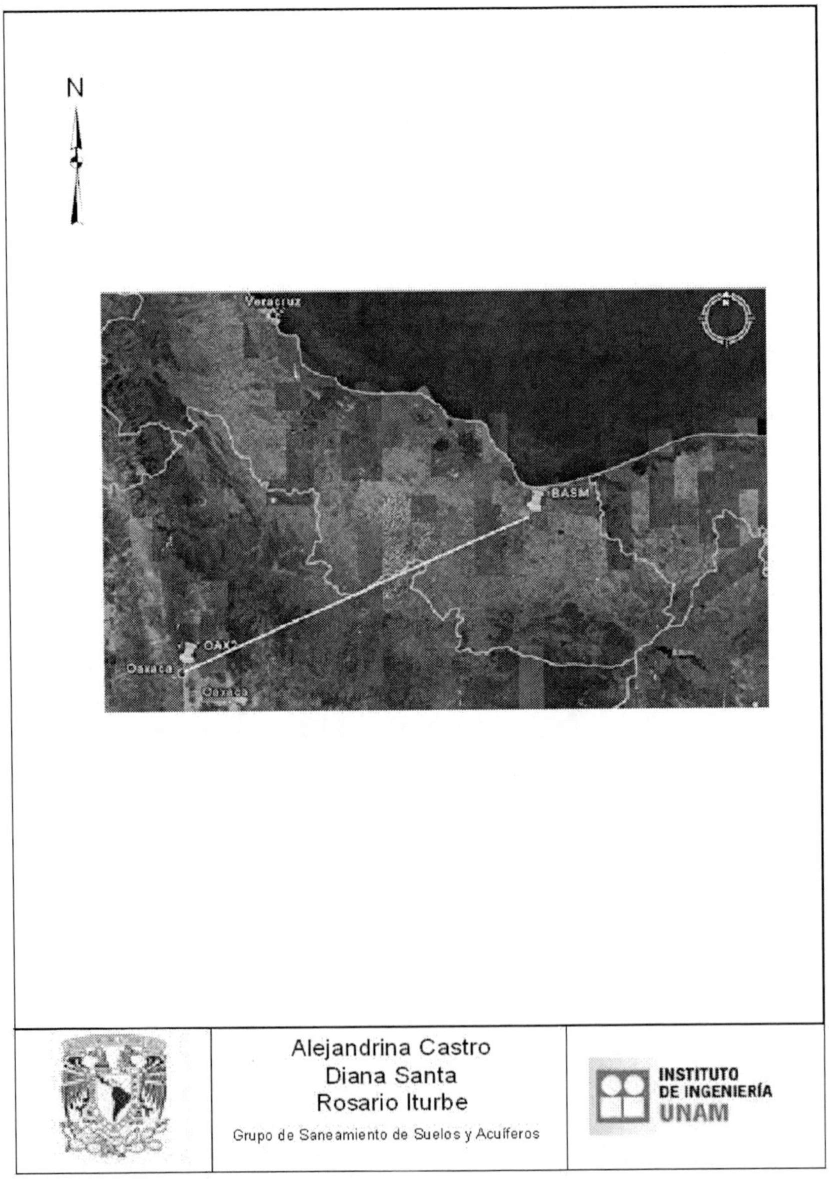

Figure 2. Base station in the study area and fixed station of RGNA, in Oaxaca state

Digital Modeling of the Terrain

The information obtained with the differential GPS was used to elaborate the Digital Model of the Terrain (Figure 3) using the Topo to Raster tool of ArcGIS, which uses an iterative technique of interpolating finite differences to create Digital Models of Terrain Elevation. This procedure can be accomplished with diverse vector-type input data (Betancur and Quintero, 2008). In this case, known elevation points obtained with the differential GPS were available and were subjected to an automatic interpolating process.

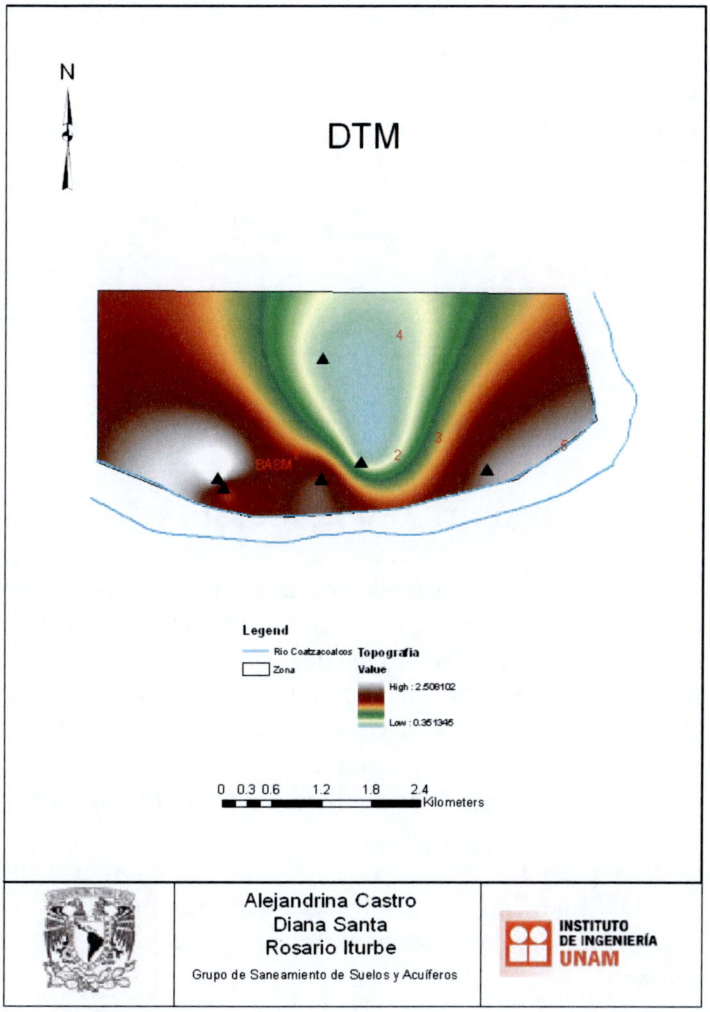

Figure 3. Digital model of the terrain (DTM)

To obtain the piezometric surface, we first used the set of field measurements of the underground water depths to perform a SPLINE-type interpolation, which is part of the Spatial Analyst tools of ArcGIS. Spline is a numerical interpolation, which uses a mathematical function that minimizes the surface curvature, yielding a soft surface that passes exactly through the input points (ESRI, 2006). This procedure originated a surface from which representative data of the phreatic level in the study area limits were extracted, complementing the original data set. This new set of data was subjected to a geostatistical analysis that allows for a correct and reliable prediction of new data by means of an exploratory analysis followed by a structural analysis. The exploratory analysis helps to identify, understand, and search for tendencies in the data, to determine the function of density, identify the type of distribution, as well as to find extreme values and variability patterns of the phenomenon by calculating the mean, mode, median, variance, bias coefficient, variation coefficient, and kurtosis coefficient. The structural analysis provides the correlations among data; it includes the adjustment of a theoretical model to the experimental semivariogram, from which the different calculation methods of the theoretical semivariogram are attained (Betancur and Quintero, 2008).

Afterwards, with the model obtained by applying Spline, we calculated the surface by means of Maps Algebra, a Spatial Analyst tool of ArcGIS, and subtracted this value from the Digital Model of the Terrain to attain the piezometric surface (Figure 4).

Table 1.

System	UTM
Zone	15 North
Datum	WGS 84
Geoidal model	EGM96

RESULTS AND DISCUSSION

Different layers were generated that allowed analyzing the information, mainly with respect to the direction of flow. Geological verification was minimal, because the whole area is reported as a flooding zone, formed by alluvial deposits.

Variation of the phreatic level between the rainy and dry seasons is approximately of 1.5 m, according to data compiled from field work during both seasons from both wells and piezometers. Acccording to metereological information, rainfall is the highest in the months from June to November, and the highest evaporation occurs in the months from April to June (EMAS, 2006).

Information on the vertical elevations is very important for hydrogeological studies, even more considering that the study area is very flat and very small differences can bring about erroneous interpretations on the direction of flow and, thereby, a wrong model regarding contaminants dispersion.

The studied swamp is located in a zone of flat topography and, hence, the direction of the underground flow is barely perceptible. This direction could not be determined with the field information, because during the rainy season, when the phreatic level is close to the surface, measurements could not be taken because the piezometers were below the water gauge With the information on the topographical elevations of the five piezometer nidi and of one piezometer, acquired with the differential GPS, we generated the digital elevation model, which provided a better view of the topography of the region (Figure 3). We used this information, because it has been demonstrated that surveys made with differential GPS yield a difference that varies from 2 to 36 mm for a base line of a maximal distance of 2,420 m

(Marín *et al.*, 2008). Figure 3 shows that the topography tends to be higher towards the south, decreasing approximately 2 m at the center of the swamp.

For the piezometric surface, the exploratory analysis suggested that no data transformation should be made; the experimental analysis, performed through the Kriging estimation geostatistical method revealed that the theoretical semivariogram calculation model that best represents the phenomenon is the exponential one.

Figure 4, which depicts the piezometric surface, reveals that the local flow direction in the east part is from east to west, whereas in the central part it is radial, with a main component toward the west. This figure also shows that in the east part, where the river changes direction drastically, there is an intrusion of the river into the swamp, this is associated with a decrease in the flow velocity of the river.

Figure 5 depicts the sampling points and some photographs taken during the field work. This figure also shows the usefulness of making hyperlinks with external documents, by means of GIS, either with texts or images, which complement the information to be shown.

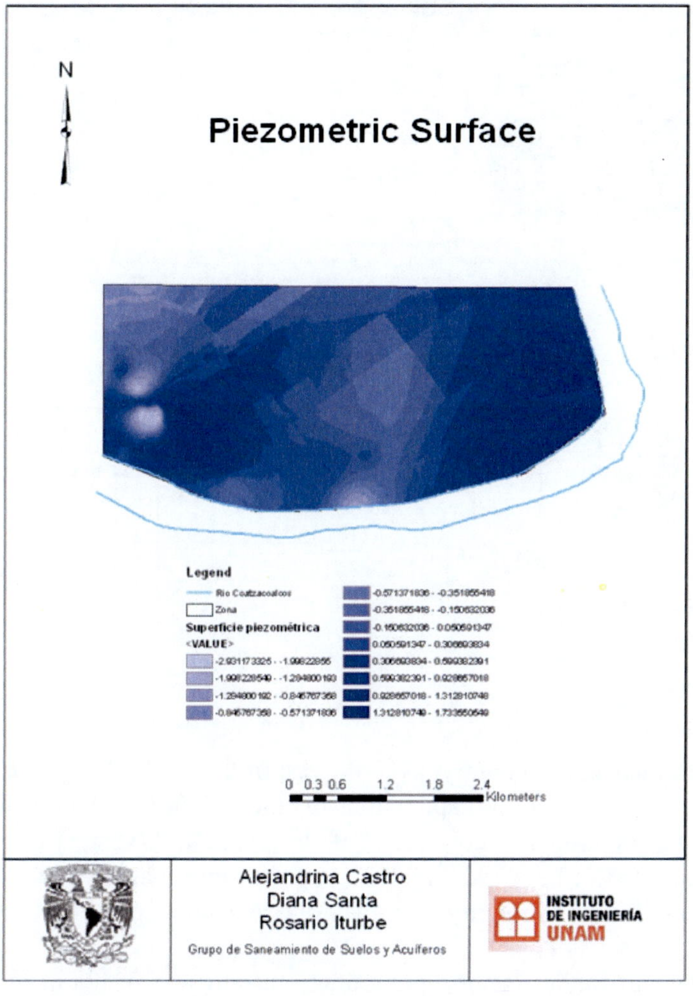

Figure 4. Piezometric surface

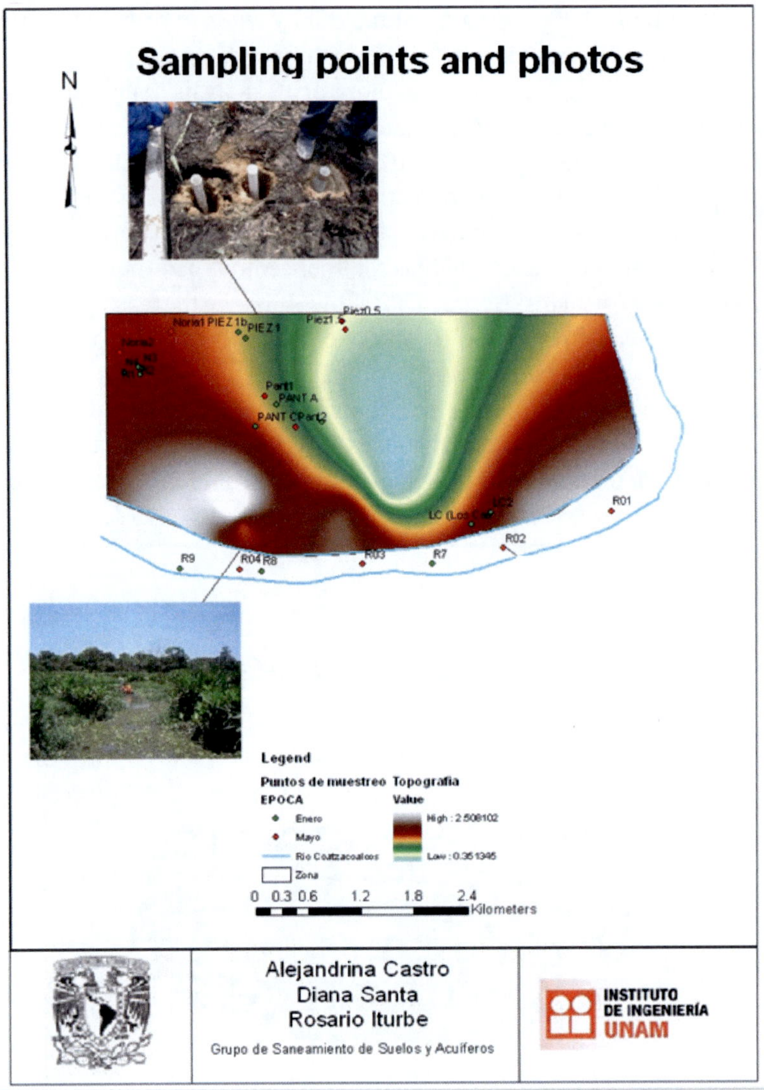

Figure 5. Sampling points

Conclusions

Underground water in our country, as in the rest of the world, has gained great relevance because of the scarcity and contamination of the surface water bodies. In Mexico, hydrogeological information on most parts of the country is very scarce, therefore, it is important that new studies be made with Geographical Information Systems that will allow monitoring the information to modify and update it, as well as making it more accessible to any user.

The use of GIS in studies lacking preliminary information or not at the desired scale is very useful, as their use allows making approximations to the reality, such as the Digital Models of the Terrain. Besides, these systems are instrumental to define important hydrogeological features, such as flow direction in areas where the conditions inherent to the site hinder their determination.

One of the application of GIS is the modeling of surfaces (piezometric surface) for which it is important to have knowledge on the behavior of the phenomenon to be studied.

Approximation to reality obtained with GIS is more precise when the field data are reliable, as provided by the information gathered with differential GPS. Likewise, if these data have a spatial distribution adequate to the extension of the study area, errors decrease and reliability increases.

Acknowledgments

Authors thank Dr. Luis G. Torres Bustillos for reviewing the manuscript and for his comments and Dr. Luis E. Marín Stillman for the suggestions made to the manuscript. We thank Geophysical Engineer Mario Balcázar González for his advice in the use of the differential GPS, José Antonio Reyna Gutiérrez for his help in collecting the information and reviewing the manuscript. We also thank Dr. Teresita Betancur Vargas and Engineer Angélica Maria Gómez for their advice.

References

[1] Barredo, J. 1996 In: Arcila, M. (2003). Sistemas de Información Geográfica y Medio Ambiente: Principios básicos. Editores: Cádiz: Servicio de publicaciones la Universidad de Cádiz. pp. 129. España

[2] Betancur, T. Quintero, D. (2008). Caracterización de la Cuenca del humedal Ciénaga Colombia, a partir de información secundaria, y utilizando herramientas geoinformáticas. Facultad de Ingeniería, Universidad de Antioquia. 2008. Colombia.

[3] CMASM (2007). Comisión Municipal de Agua y Saneamiento de Minatitlán. Informes internos de pozos. Mexico.

[4] CNA (2002). Determinación de la disponibilidad de agua en el acuífero Costera de Coatzacoalcos, Estado de Veracruz, Subdirección General Técnica, Gerencia de Aguas subterráneas., Comisión Nacional del Agua, Mexico.

[5] CONAGUA (Comisión Nacional del Agua) (2008). Atender la problemática de sobreexplotación de acuíferos, uno de los retos más importantes del país. Coordinación de comunicación de CONAGUA, Subgerencia de información, comunicado de prensa No. 008-08. Mexico, D.F. January 22, 2008. http://www.conagua.gob.mx

[6] DOF (2001). Diario Oficial de la Federación. [Decree that establishes and makes known to the general public the single denomination of the aquifers recognized in the Mexican United States by the National Water Commission, and the homologation of names of aquifers used for the emission of concession titles, assignation, or permits granted by this decentralized agency]. December 5, 2001.

[7] ESRI (2006). Accessed on July 10, 2008. Spline. www.esri.com

[8] EMAS (2006). Estaciones Meteorológicas Automatizadas. Servicio Meteorológico Nacional. Comisión Nacional del Agua, Mexico.

[9] INEGI (2000). Información Geográfica, hacia el tercer milenio. Instituto Nacional de Estadística, Geografía e Informática. Aguascalientes, Mexico.

[10] INEGI (2007). Accessed July 20, 2008.. www.inegi.gob.mx Instituto Nacional de Estadística, Geografía e Informática. Information on hydrology.

[11] INEGI (2008). Sistemas de Información Geográfica. Personal information provided by the institution.

[12] Lindig-Cisneros, R. and Zedler, J. (2005). La restauración de Humedales. Temas sobre restauración ecológica. Secretaría de Medio Ambiente y Recursos Naturales. Instituto Nacional de Ecología. U.S. Fish and Wildlife Service. Unidos para la Conservación, A.C. , Mexico, D.F, MEXICO

[13] Lyon, J. and McCarthy, J. (1995). Wetland and Environmental Applications of GIS. Lewis Publishers. USA

[14] Marín, Luis Ernesto (2004). El Agua en México: Retos y oportunidades. Revista Real Academia de Ciencias Exactas, Físicas y Naturales. Vol. 98, No. 2. page. 287

[15] Marín, L. E., Balcazar, M., Ortiz, M., Steinich, B. and Hernández-Espriu, J.A. (2008). Comparison of elevation heights using a Differential Global Positioning System (DGPS) and a Total Station. Geofísica Internacional 47 (1), 81-83.

[16] Marín, L.E. (2008). Personal communication.

[17] Moya, B., Hernández, A. and Elizalde, H. (2005). Los humedales ante el cambio climático. Investigaciones Geográficas, 37. 127-132.

[18] Salas, M.A. (2007). Aplicación y utilidad de Sistemas de Información Geográfica en proyectos de hidrogeología. Memorias de la Convención Nacional de Geografía 2007. Guadalajara, Jalisco, Mexico. April 18 to 20, 2007.

[19] Vega, G. (2001). El papel de la cartografía digital en el funcionamiento de los Sistemas de Información Geográfica. College Thesis. Universidad Nacional Autónoma de México. Facultad de Filosofía y Letras. Colegio de Geografía. pp. 154, Mexico.

[20] Yañez-Trujillo, L, Mojena-López, E., Martínez-Méndez, L., Montes de Oca, A., Bastida-Villalobos, V., Gallegos-Martínez, M. and Gutiérrrez-Rojas, M. (2003). Estrategias para la evaluación de impactos petroleros por medio de percepción remota y sistemas de información geográfica en México. Mapping International. Revista Internacional de Ciencias de la Tierra. Web page accessed on July 28, 2008. www.mapping.interactivo.com.

In: Remediation of Soils and Aquifers
Editors: Luis G. Torres and Erick R. Bandala

ISBN 978-1-60741-372-1

Chapter 5

NATURAL ATTENUATION OF CONTAMINATED SITES

Luis Carlos Fernández Linares
Unidad Profesional Interdisciplinaria de Biotecnología-IPN

1. INTRODUCTION

Contamination of sites by all sorts of organic and inorganic contaminants, from natural and synthetic sources, has become an increasing problem in the last decades due to industrialization. Some of the main contamination sources are due to the uncontrolled, indiscriminated and, many times, careless discharges; to mismanagement of wastewaters, hazardous wastes, and inadequate disposal of wastes; to accidents during handling and distribution of hazardous substances, such as tanker or road spills, and clandestine poaching of oil derivatives. The large amount of contaminated sites has led to the need of adequate techniques for their recovery, being economy an important factor. It is estimated that in the USA there are some 330 000 contaminated sites, with a remediation cost of hundreds of million dollars (U.S. EPA, 1999), in Europe, 750 000 sites are suspected to be contaminated due to human activities; a soil recovery cost of ca. 93 000 million Euros has been estimated for eight countries of the European Community (European Environment Agency). Bento *et al.* (2005) provide a complete inventory of contaminated sites from diverse industrialized countries. Regarding oceans, it is calculated that they become contaminated with 3.2 million tons of hydrocarbons each year (Heller, 1993). It can be inferred that the amount needed to recover these sites would reach several million dollars.

There are different physical, chemical and thermal technologies available to treat contaminated soils, these include excavation and confinement, vapor extraction, stabilization and solidification, soil washings, burning, chemical precipitation, vitrification, thermal desorption, and incineration. Due to their high cost, biological processes have been pointed out as more attractive clean-up technologies of contaminated soils, because of the feasibility of reducing cost, partial or total elimination of the contaminant, together with a reduction in the associated risk and greater social acceptance (Baker & Herson, 1994)

Currently biological methods are used in 15 to 20% of the cases, presenting economical and technical advantages, lower risks, being less aggressive to the environments and having reduced energy requirements as compared with the physical, chemical, and thermal technologies. Biological processes are the only methods, except for incineration, that achieve mineralization of organic contaminants, releasing innocuous products. However, it requires more time. Currently, a wide number of bioremediation techniques are known, both *ex-situ* and *in-situ*. The former include land-farming, biopiles, bioreactors, and composting, whereas biostimulation, bioaugmentation, bioventing, and phytoremediation are applied both *in situ* and *ex situ*. In the last years, the use of technologies based on their risk on human health have arisen, hence, Natural Attenuation (NA), also known as Passive or Intrinsic Bioremediation have come into being as an option for site remediation. In the USA it has become an important remediation technology (NATO/CCMS, 2003), in 1997 it was applied at 15 000 sites (Renner, 2000).

2. Definition of Natural Attenuation

There are different definitions of *Natural Attenuation (NA) or Monitoring Natural Attenuation (MNA*), some of them are dealt with in this chapter; likewise, in the last years new proposals or variants from the basic process have arisen, such as the *Enhanced Natural Attenuation*, which in a strict sense of the process, are active bioremediation technologies (with human intervention), defined as biostimulation, bioaugmentation, bioventing or phytoremediation. Processes aimed at incrementing the natural processes by adding or varying a factor of the system.

Several agencies have developed definitions on natural attenuation, four of them were chosen that describe adequately the process. The United States Environmental Protection Agency (US EPA) uses the term of supervised or monitored natural attenuation, placing emphasis on the follow-up, supervision or monitoring of the process, and defines it as: "reliance on natural attenuation processes (within the context of a carefully controlled and monitored site cleanup approach) to achieve site-specific remediation objectives within a time frame that is reasonable compared to that offered by other more active methods. The "natural attenuation processes" that are at work in such a remediation approach include a variety of physical, chemical, or biological processes that, under favorable conditions, act without human intervention to reduce the mass, toxicity, mobility, volume, or concentration of contaminants in soil or groundwater. These *in-situ* processes include biodegradation; dispersion; dilution; sorption; volatilization; radioactive decay; and chemical or biological stabilization, transformation, or destruction of contaminants." (National Research Council).

The European Union for Responsible Incineration and Treatment of Special Waste (Erutis) refers to it as: "the combination of all natural degradation processes that take place in soil and groundwater, such as biodegradation, adsorption, dilution and chemical reactions with the soil itself. Over time, these natural processes reduce the concentration of contaminants to acceptable levels, in other words, levels that pose no risk for human health and for the health of the ecosystem. Free-floating, or mobile, pollutants can spread in liquid or vapour form and are thus easily captured by degradation processes. On the other hand, organic compounds that have a high molecular weight and many inorganic substances become trapped in the soil, which impedes their diffusion".

The *U.S. Air Force Center for Environmental Excellence* defines it as: "from the integration of several subsurface attenuation mechanisms that are classified as either destructive or nondestructive. Biodegradation is the most important destructive attenuation mechanism. Nondestructive attenuation mechanisms include sorption, dispersion, dilution from recharge, and volatilization".

The U.S. Army defines it as: "the reduction of contaminant concentrations in the environment through biological processes (aerobic and anaerobic biodegradation, plant and animal uptake), physical phenomena (advection, dispersion, dilution, diffusion, volatilization, sorption/desorption), and chemical reactions (ion exchange, complexation, abiotic transformation). Terms such as intrinsic remediation or bio-transformation are included within the more general natural attenuation definition" (U.S. Army, 1995).

Natural attenuation as a strategy requires assessment of the health and environmental risks involved; a hydrological, physicochemical and biological characterization of the contaminated site, aside from and adequate and strict monitoring of the main involved parameters. It is important to point out that the selection of application of Natural Attenuation is not synonymous to an alternative without actions; it is necessary to evaluate if the conditions exist for its application, to follow-up and document how it is being accomplished; both characterization and monitoring are utmost important.

Natural attenuation must reduce the potential risk in three ways: 1) transforming the contaminant into a less toxic compound through degrading processes such as biodegradation or abiotic transformation, 2) reducing the concentration of the contaminant, by which exposure potentials can be abated, and 3) diminishing the mobility and bioavailability of the contaminant through its adsorption to the soil and rocky matrix.

NATURAL ATTENUATION PRESENTS THE FOLLOWING ADVANTAGES AND LIMITATIONS

- Contaminants are transformed *in situ* to less toxic compounds, i.e., CO_2 and H_2O; which is not accomplished with the physicochemical technologies.
- Natural attenuation allows recovering a high percentage of the natural features of the soil, in contrast to other technologies that alter the original structure of the soil; hence, it is considered a highly sustainable technique.
- Contaminants are transferred to another phase.
- Flexible application to the whole or part of the site.
- Applicable after an active remediation procedure.
- Does not require equipment for the treatment or disposal of wastes.
- Contaminants and/or soil are not transported to other sites, which could represent a risk for the potential receptor.
- It is cheaper, costs are only needed for monitoring procedures.
- Can be integrated as part of a general clean-up operation for highly contaminated sites.
- Is a process that can be slow.

- Requires a complete characterization of the site, hence, monitoring costs are higher than those of monitoring during an active clean-up.
- Efficacy of natural attenuation can be affected by changes in the environment (climatological) or induced ones (engineering).
- It is necessary to inform better the society about this technology.

It is important to keep in mind that natural attenuation does not require investment for the process, but it does require a complete and effective characterization of the site, as well as of a detailed analysis of the soil's prevailing conditions. This implies an investment for collection and analysis of samples, installation of monitoring wells and for the analytical or numerical simulation of the attenuation, using specific information of all or part of the following parameters:

- Source of the contaminant and amount of contaminants present.
- Flow of underground currents.
- Distribution of the contaminating phase.
- Partition and concentration of the contaminant and of the electron acceptors (dissolved oxygen, NO_3, SO_4, Fe(III), Mn (IV), CO_2) in the different phases of the soil (solid, liquid, and gaseous).
- Identification of the native microbial population.
- Rate of biological and non-biological degradation.
- Hydraulic characteristics of the phreatic mantles, for example, whether dealing with an open or a confined mantle or, in its case, of the saturated zone of the soil.
- Physicochemical properties of the soil.
- Seasonal variations: level of groundwater or seasonal changes in pluvial precipitation (rainfalls).

3. Processes Participating in Natural Attenuation of Compounds

The processes that participate in attenuation are physical, chemical, and biological that, which under favorable conditions, act without human intervention, reducing toxicity, mobility, volume. or concentration of the contaminants in the soil or the groundwater. Among these processes are dispersion, dilution, sorption (absorption and desorption), volatilization, photooxidation, as well as degradation, stabilization, transformation or chemical and biological destruction of contaminants (Fig. 1).

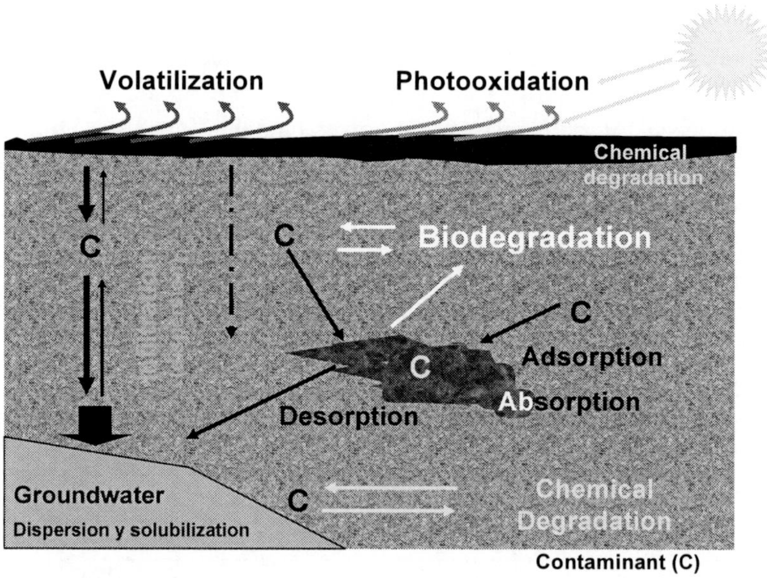

Figure 1. Processes participating in natural attenuation of contaminants.

3.1. Biodegradation

This is the most important factors that participates in eliminating contaminants from the environment and, consequently, in natural attenuation, and is defined as the process by which substances are broken down by living organisms, mainly microorganisms, that transform organic molecules into fatty acids, carbon dioxide, water, and biomass.

Biodegradation is one of main processes because a large number of organic compounds, such as petroleum hydrocarbons, BTEX (benzene, toluene, ethylbenzene, and xylene), polyaromatic hydrocarbons, pesticides, aliphatic and aromatic hydrocarbons, chlorinated compounds, pesticides and nitrate aromatic hydrocarbons (explosives) can be transformed or degraded by microorganisms. Biodegradation of contaminants in a site depends on different factors: the microflora present; nutrients concentration (N and P), oxygen or electron acceptors concentration, temperature, salinity, as well as on the chemical structure of the hydrocarbon. Some compounds are recalcitrant (poorly degradable) such as some of the heavy, polyaromatic, or halogenated hydrocarbons.

Biodegradation can be accomplished in both aerobic and anaerobic conditions; however, the aerobic metabolic pathway is more efficient and biodegradation of contaminants is achieved faster. Microorganisms can use a large variety of electron acceptors, these can be O_2, NO_3, NO_2, $(SO_4)^{-2}$, Fe(III), and Mn (IV). Oxygen is an important acceptor because of its distribution in the environment and its participation in reactions, when it is used as acceptor it is an aerobic process. The use of the remainder compounds or metals as acceptors is called anaerobic metabolism. Hence, the microbial processes that can be performed, in function of the redox state, are: aerobic respiration, denitrification, sulfate, iron, and manganese reduction, methanogenesis, hydrogen and carbohydrates oxidation (Table 1). Each of these pathways generates a final product that can be determined at the site as an indication that the biodegradation process (CO_2, H^+, H_2S, CH_4, Cl^-) is being performed.

During assessment of natural attenuation of a contaminated soil, redox processes must be taken into account, some of which, such as Mn and Fe reduction, have been poorly studied (Ferrey *et al.*, 1998). Reaction with a high enthalpy yield, as are aerobic respiration and denitrification, proceed toward processes of lower enthalpy, such as Mn, Fe, and $(SO_4)^{-2}$ reduction (Table 1). Biological catalyzed reactions, as those occurring in the soil after it has been exposed to an organic contaminant, follow this rule, giving rise to zones with different redox potential (Fig. 2).

Table 1. Redox processes occurring during a natural attenuation process.

Process	Reaction	ΔG° [KJ]	Eh [mV]
Aerobic respiration	$CH_2O + O_2 \rightarrow CO_2 + H_2O$	-502.3	>300
Denitrification	$CH_2O + \frac{4}{5}NO_3^- + \frac{4}{5}H^+ \rightarrow CO_2 + \frac{2}{5}N_2 + \frac{7}{5}H_2O$	-476.8	>300
Mn (IV) reduction	$CH_2O + 2MnO_2 + 4H^+ \rightarrow CO_2 + 2Mn^{2+} + 3H_2O$	-340.3	300-100
Fe(III) reduction	$CH_2O + 4Fe(OH)_3 + 4H^+ \rightarrow CO_2 + 4Fe^{2+} + 11H_2O$	-115.0	300-100
SO_4^{2-} reduction	$CH_2O + \frac{1}{2}SO_4^{2-} + \frac{1}{2}H^+ \rightarrow CO_2 + \frac{1}{2}HS^- + H_2O$	-104.7	<100
Formation of CH_4-	$CH_2O \rightarrow \frac{1}{2}CH_4 + \frac{1}{2}CO_2$	-92.9	<-100

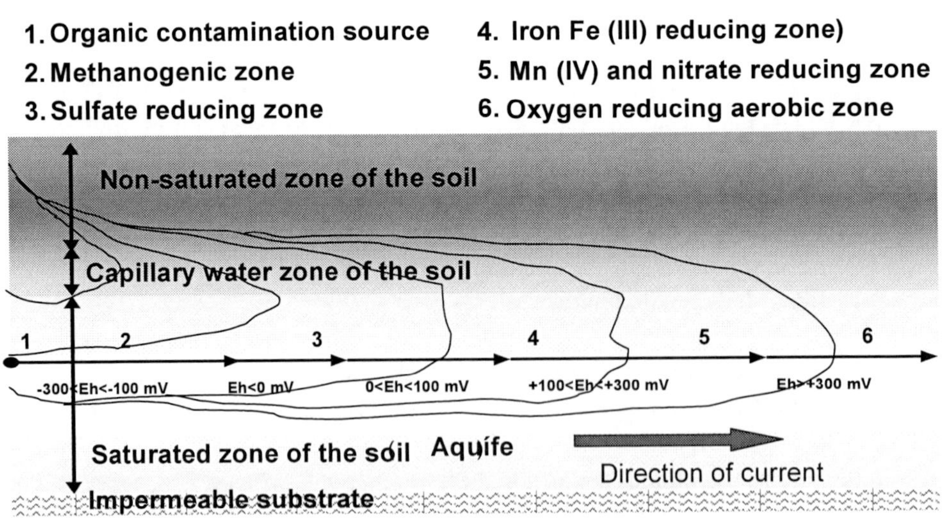

Figure 2. Variation of redox potential in different zones of a contaminated site.

3.2. Biotransformation

Some inorganic compounds, such as metals and non-metallic ones (As, Cr. Hg), although not biodegradable, they can become biotransformed; microorganisms can change the oxidation state of metals (Wang & Mulligan, 2005), precipitate them, as well as mediate their methylation or demethylation, making them less toxic and/or available, decreasing human health and environmental risks.

3.3. Sorption

Contaminants in function of their physical and chemical properties (polarity, charge, size, chemical structure, and oxidation state) can be absorbed or adsorbed by soil particles, reducing their movement or transport (delaying) or retaining (immobilization), reducing the human and environmental risks. Sorption can influence volatilization and biodegradation by making them less available. This process tends to be reversible, inducing dispersion and partition of molecules. In the case of hydrocarbons, because of their apolar behavior, they are commonly adsorbed by hydrophobic linkages.

3.4. Chemical Transformation

Some contaminants are degraded by chemical reactions that occur naturally in the soil (i.e., not mediated or facilitated by organisms). Redox-sensitive surfaces, such as Fe and Mn (hydro)oxides, organic matter, mineral clay, have shown the ability to oxidize compounds such as As. The most important types of reactions occurring in the soil are oxide-reduction, acid base, and hydrolysis (Eweis, *et al.*, 1999). However, most hydrocarbons are not significantly degraded by chemical reactions. It is important to point out that addition of oxidizing agents, such as potassium permanganate, to the soil or aquifer is not natural attenuation.

3.5. Volatilization

Several organic compounds, mainly those of low molecular weight, can evaporate in the atmosphere, where they are dispersed by the air currents, reducing their concentration in the site; the compound in the atmosphere can be photooxidized, which contributes to its degradation, or it can be moved to another site, or diluted. Vapors in contact with the soil can be biodegraded easier by the existing microflora. Vaporization occurs mainly in recent contaminations and is an important factor to be considered as an exposure pathway in a risk analysis.

3.6. Photooxidation

Photooxidation, an oxidation reaction induced by light (IUPAC, 1997), is an important pathway for hydrocarbons transformation, mainly in marine environments (Laane *et al.*, 1985). Aromatic hydrocarbons are more sensitive to photochemical degradation than the saturated ones (Taghvaei *et al.*, 2007). Photodegradation can give rise to more soluble (Jacqout *et al.*, 1999) and more toxic (Lacaze & Villedon De Naide, 1976, Ostgaard *et al.*, 1987) compounds.

4. Processes and Mechanisms Involved in Natural Attenuation

A fundamental part of natural attenuation is the interaction contaminant-soil-water, where mass transfer by physical and chemical adsorption is accomplished, depending on the nature, charge, and reactive properties of the soil and contaminant particles. In the soil, mineral particles, such as kaolinites, ilitites, and chloritites, among others, together with the functional groups of the inorganic fractions of the soil generate charges, both positive and negative, in the surface of the soil particles. On the other hand, iron, aluminum, manganese, titanium, and silica oxides are found in the soil, whose charge depends on the basic structure and the medium's pH. Likewise, there are also the different functional groups (OH, CO_3, CN, NH, NO_2) of the organic matter found in the soil, which can protonate or deprotonate as a function of the pH. The aforementioned factors and properties influence the mobility and reactivity of contaminants.

The reactive surfaces of the soil fraction or the solid soil only provide the location for the reaction between the soil and the contaminants found in the water of the soil pores. The nature of the binding force, between the contaminant and the soil, depends on the functional groups of their surfaces. Together with the characteristics of the contaminant and the soil, within this system, there is also the influence of the water retention capacity, moisture and energetic interaction of the water with the soil (polarity, capillarity, and osmosis). In the pore, containing the water and the contaminants, acid-base, hydrolysis, and oxide-reduction type chemical reactions occur that depend on the functional groups of the three factors (soil, water, contaminant), on the salts and the pH; these reactions can be mediated by biological activity. The physical and chemical properties of the soils influence importantly aeration, availability of nutrients, salts, and metals, as well as water retention, therefore they exert a considerable effect on the biological activity.

4.1. Mobility and Solutes Partition

Movement of contaminants in the soil is carried out through advection, dispersion, and transfer between phase mechanisms. In function of their nature, contaminants can be transported at different velocities and become chemically transformed while advancing through the soil, heterogeneity of the soil (contents of clay, lime, and sand) also influences water and contaminants movement. Chemical compounds and ions move with the fluid because of advection (movement of particles or of a contaminant due to the velocity of the flow), by the hydrodynamic dispersion process that occurs when there is a concentration gradient giving rise to contaminants movement by mechanical dispersion and molecular diffusion. Partition of contaminants occurs when the molecule stops moving by advection or dispersion and is retained or eliminated by the adsorption, dissolution, absorption, and volatilization processes previously described, except the dissolution process that refers to the amount of pure contaminant that will become dissolved in water, and is limited by the solubility of the compound. When a compound is in its pure form, its concentration in water will depend on the solubility and mean concentration (mg/L), (Eweis, *et al.*, 1999).

4.2. Mobility and Metals Partition

A special case of natural attenuation are the heavy metals, which cannot be degraded. To insure that the metals that are in the aqueous phase of soil do not represent a hazard anymore, they must undergo one of the following processes, complexing, precipitation, and sorption by the soil (Maguire *et al.*, 1981; Young et al., 1990; Yong, 2000). However, one must be very careful, since some processes are reversible because mobility and bioavailability of metals depend on factors, such as hydrogeology of the site, changes in acidity, ionic strength, as well as redox potential of the system and formation of complexes.

Partition of metals, that is, their distribution between the soil (immobilized) and the water (available) of a site is an important variable to be determined for natural attenuation. Determination of metals partition must be developed experimentally by means of two assays: 1) absorption isotherms in batch equilibrium and 2) in columns (lysimeters) or leachate cells, both described by Yong (2000)

5. Monitoring and Sampling for Natural Attenuation I

Monitoring or follow-up of the remediation processes, be them physical, thermal, or biological, is an essential factor of any treatment or response to a contaminated site. Monitoring is performed to assess the efficiency of the processes applied to the recovery of the site aimed at warranting the protection to human health and the environment, and compliance with the standards established by each country or state (contamination limits, or based on risk). For natural attenuation, monitoring or follow-up is particularly important, due to the nature of the process, that can lead to the decision of not carrying out an action; however, it is a process that must be supervised or monitored to ensure a reduction in contaminants, of the migration potential, and of the risk. In this way, Natural Attenuation started to be named *Monitored Natural Attenuation, MNA*, making emphasis on the need and relevance of follow-up and monitoring of the processes being carried out in the soil and aquifer.

For the monitoring and follow-up of the natural attenuation of a site, a protocol must be set forth, which is specific for each site. It is necessary to establish the wells and sampling points of the site; for a contamination plume. it will be necessary to include at least one well at the initial part of the contamination source, two or more in the plume, and another in an area free of contamination in the direction of the flow. In the case of soils, it is convenient to localize wells around the site or at least on both sides of the site in relation to the direction of the flow (Fig. 2). A well must be placed between the contaminated site and its potential receptor to estimate the risk for human health. The number of sampling points of the soil must be determined statistically and must be representative; there are guidelines and standard methods from ASTM (1995, 1996, and 1998) and EPA (1991, 1992, 1993; 1994a, and 1995) oriented to hydrocarbon contaminated soils (API, 1996; PROFEPA, 2000) and of the Environmental Mexican Norm (DOF, 2003). The rate of natural attenuation of a site can be determined by three approximations: a balance of mass, concentration vs. time, and concentration vs. distance.

To determine where NA is an appropriate solution for a soil or an aquifer in a given site, the following must be considered:

1. That the contaminants present in the soil or aquifer can actually be eliminated by the natural attenuation processes.
2. Whether the contamination plume is stable or not, and the potential of the environmental condition to influence the stability of the plume and the changes in time.
3. Whether human health, sources of drinking water, other groundwater and surface water, ecosystems, sediments, air, or other environmental resources can be affected by choosing natural attenuation.
4. Whether the contaminant by itself, or as an accumulation with other neighboring sources, can exert a long-term impact on the environmental resources.
5. If the time estimated for the remediation is reasonable as compared with the times required by other active methods.
6. The nature and distribution of the contamination source and whether it is or can be controlled.
7. If the products resulting from the transformation present a higher risk, because of an increase in their toxicity or mobility.

In the case of heavy metals that are adsorbed in the soil, monitoring must include analyses of soil and water, determining the distribution, concentration, and speciation of the heavy metals, pH and Eh of the medium.

The monitoring program for each site must specify the location, frequency, and type of necessary samples and measurements to assess NA. Monitoring must comply with the following:

- Demonstrate that NA is occurring according to expectations.
- Detect changes in environmental conditions (hydrological, microbiological, etc.) that might reduce the efficiency of the process.
- Identify any product transformation that could increase its toxic potential and mobility.
- Verify that the plume will not expand.
- Verify that no impact will occur on other receptors.

- Detect new releases of contaminants to the environments that might impact the NA's efficiency.
- Verify how much the marked objectives have accomplished.

The parameters to be determined in monitoring natural attenuation processes will depend on each site and type of contaminants; it must include the physical, chemical, and biological aspects of the process. In general, the parameters to be determined are (Table 2):

Table 2. Parameters to determine on the soil and water samples, with the respective methods for the follow-up of natural attenuation.

Analysis	Field or laboratory	Method
Total Petroleum hydrocarbons	Laboratory	EPA 418
BTEX, MTBE and TAME	Laboratory	EPA 8021
Total hydrocarbons, Diesel grade,	Laboratory	EPA 8015M
Polyaromatic hydrocarbons	Laboratory	EPA 8310 (17 HPA)
Electrical conductivity	Laboratory	EPA 120.1 o SW-9050
pH	Laboratory	EPA 150.1 o SW-846, 9045 / SW 846, 9040
Temperature	Laboratory	EPA 170.1
Dissolved oxygen	Laboratory	SM 4500C / SM 4500G
Leachable Cr VI ions	Laboratory	Digestion, lixiviation and EPA 7196A
Leachable metals (As, Ba, Cd, Hg, Ni, Ag, Pb, and Se)	Laboratory	Digestion, lixiviation and EPA 6010A
Texture	Laboratory	Bouyoucus' hydrometer
Moisture	Laboratory	Gravimetric
Base saturation percentage (K+, Ca++, Mg+++, Na +)	Laboratory	Extraction with ammonium acetate and quantification with AA
Cationic exchange capacity	Laboratory	Washing with Alcohol/Destilation/Neutralization
Total organic carbon	Laboratory	Automatic Carbon autoanalyzer
Total carbon	Laboratory	Automatic Carbon autoanalyzer
Total nitrogen	Laboratory	Kjeldahl
Available phosphorus	Laboratory	Olsen (colorimetric)
Carbonates	Laboratory	Digestion with HCl and volumetry
Nitrates, ammonium and nitrites	Laboratory	Extraction with KCl, vapor carrier and volumetry
Sulfates	Laboratory	Extraction with $BaCl_2$ and turbidity measure
Water soluble phosphates	Laboratory	Colorimetry
Total phosphorus	Laboratory	ASTM D 1091
Total Metals: Al, As, Ba Cd, Cr, Cu, Fe, Hg Ni, Pb, Se, V, and Zn	Laboratory	SM 3113B / SW 846 6010 / or EPA 200.7 / EPA 200.8 / EPA 200.9
Lead	Laboratory	EPA 7420 y 7421

Table 2. Continued.

Analysis	Field or laboratory	Method
Nitrate	Laboratory Field	EPA 35302 SW-9056 Calorimetric Kit
Sulfate	Laboratory Campo	EPA 375.4 o SW-9056 Calorimetric kit
Manganese	Laboratory Field	EPA 3010/200 Calorimetric kit
Methane	Laboratory Field	Head Space; CG/FID Portable equipment
Carbon dioxide	Laboratory Field	Head Space; CG/TCD Portable equipment
Sulfhydric acid	Laboratory Field	ASM D 4658 Portable equipment
Aerobe microbiology: hydrocarbonoclast bacteria; total fungi; and hydrocarbonoclast fungi	Laboratory	Plate counting or most probable number
Anaerobe Microbiology: Fermentative, Sulfate-reducing, iron-reducing, nitrate-reducing and methanogenic bacteria	Laboratory	Most probable number

Sampling of soil and water, storage and conservation of samples must be performed according to established standards or norms, implementing custody chains.

5.1. Biological Indicators

To demonstrate biological participation in eliminating organic contaminants during natural attenuation, some of the aforementioned analyses are required, which as a whole allow inferring biodegradation. Likewise, it is possible to establish biodegradation tests at the laboratory level. During biodegradation, products such as CO_2, CH_4 and H_2S are generated and there is consumption of electron acceptors in the aerobic O_2 process, as well as of NO_3, H_2, Mn^{4+}, Fe^{2+}, SO_4^{-2} and CO_2 during anaerobiosis. Field determination of these parameters together with the mass balance (loss of contaminant), and additionally the laboratory tests, permit to ensure that loss of the organic contaminant is through a biological pathway. Other techniques, such as determination of enzymatic activity (Serrano *et al.*, 2007), biomarkers (Atteia, & Guillota, 2007, Nijenhuis *et al.*, 2007), isotopes, and molecular tools (Show and Hicks, 2005; Davis et al., 2002) have been applied to study NA.

5.2. Risk Assessment

Risk assessment can have different goals; it can have an industrial, civil, or environmental focus. Risk assessment does not only determine the effect of a contaminated site, but can also establish contamination limits, prevent contamination with new chemical and pesticide products, set forth the reliability of engineering in new industrial and technological activities, implement evaluation of the environmental and financial impact, or establish clean-up measures. In the context of contamination, risk assessment has been used in Europe mainly to prioritize sites treatment, that is, as an indicator with comparative goals. Risk analysis will probably permit to determine the feasibility of applying the natural attenuation process as a technology to recover a given site. Risk analysis could also be used as another parameter to be determined, aimed at decreasing the risk.

Risk can be defined in different ways, in the realm of environmental contamination, it could be said that it is the probability that, in a given situation, a hazardous substance might produce damage (Peña, *et al.*, 2001.); and risk assessment as the process to estimate the probability that the damage will occur in humans or the ecosystem as a consequence of their exposure to toxic substances. For it, the hazard must be identified as well as the dose-response relation, and exposure and risk characterization must be accomplished. The subject is ample and out of the reach of this chapter, hence a review of the following literature is recommended.

Risk assessment of chemicals. An introduction (1995) (Van Leuwen, C Jy Hermens, JLM)

Risk based decision making for assessing petroleum impact at exploration and production sites (2001) (McMillen, S. et al.)

Hazardous Waste Management (2001) (LaGrega, M D, Buckingham, PL y Evans. J C).

Standard Guide for Risk-Based Corrective Actions Applied at Petroleum Release Sites (ASTM, 1996)

http://www.defra.gov.uk/corporate/busplan/riskmanage/index.htm Department for Environment, Food and Rural Affairs

6. Study Cases and Application of Natural Attenuation

There are different studies and reports on the application of natural attenuation to diverse contaminants, from metals to pesticides, going over diverse types of hydrocarbons. Table 3 compiles some natural attenuation cases at the field and laboratory levels.

Table 3. Cases of application of natural attenuation

Compound	Field-Laboratory	Site	Reference
Arsenic	Field	Bangladesh, India	Stollenwerk *et al.*, 2007
Arsenic	Field	Tasmania	Gault *et al.*, 2005
Benzene and dichloromethane	Field	Michigan USA	Ravi *et al.*, 1998
BTEX	Field, 42 sites	USA	US Aire Force
BTEX	Field	France	Atteia and Guillot, 2007
BTEX, TPHs	Field	USA	MAFRTR, 2000 y 2001
Petroleum	Field	Antarctica (sea)	Snape *et al.*, 2005

Table 3. Continued.

Compound	Field-Laboratory	Site	Reference
Diesel	Laboratory	USA and China	Bento *et al.,* 2005
Diesel	Laboratory	Spain	Serrano *et al.*, 2005
Chlorinated ethenes	Field	USA	Davis *et al*., 2002
Gasoline	Field	USA	Khan & Husain, 2001
Gasoline	Field	USA	MAFRTR, 2001
Gasoline	Field	USA	Dupont *et al*., 1998
Aliphatic chloride hydrocarbons	Field, 14 sties	USA	MAFRTR, 2004
Metals, As, Pb, Zn, Cu.	Laboratory	Korea	Pyeong-Koo *et al.*, 2005
MTBE and Benzene	Field	Canada	Schirmer *et al.*, 1998
MTBE, BTEX	Field	Taiwan	Chen, *et al*., 2005
Petroleum	Field (sea and beach)	Galicia, Spain,	Gallego *et al*., 2007
Petroleum in soil	Field	USA	MAFRTR, 2004
TCE	Field	Connecticut USA	Chapman *et al.*, 2007
TCE	Field and laboratory	Oklahoma, USA	Kennedy *et al.*, 2005
TCE	Field	Germany	Nijehuis *et al.*, 2005
TCE	Field	Tennesse, USA	Lenczewski *et al.*, 2003

6.1. Case Study of Natural Attenuation in Mexico

There is no report at all on the application of natural attenuation for the treatment of either groundwater or soils in Mexico. Up to this date, it is not contemplated in the norms, neither are there any national parameters, guidelines, or protocols, nor any institutions or enterprises taking NA into account in their policies. NA or MNA has been applied or developed mainly in the US by EPA, probably addressed at groundwater contaminated with chlorine compounds, VOCs, and BETX. The problematic and realities of Mexico are different; one of the main problems consists of the soils contaminated with petroleum and its derivatives; another problem are metals, which are hard to be treated through NA. At the experimental level, the Mexican Institute of Petroleum (IMP, for its initials in Spanish) developed a research project addressed to a specific problem in Mexico, soils contaminated with hydrocarbons, particularly with perforation wastes (Fernández *et al*., 2003) .

The project was carried out at two sites during three years, monitoring natural attenuation, performing ecotoxicological and risk assessments; biodegradability tests in the laboratory, and in a complementary way, studies were performed with a lysimeter (NA and biostimulation) as well as active treatment of biopiles. That work provided the guidelines to establish the feasibility and follow up of natural attenuation mechanisms in perforation waste deposits.

During the three years of the study and NA follow-up, in both study sites a diminution of contaminants was found (Table 4), that can be attributed to the microbial activity, both aerobic and anaerobic, that there is no risk to human health, as well as a decrease in ecotoxicity, which demonstrates a self-cleaning process of the sites. Not withstanding on the surface areas, where the highest concentration of hydrocarbons (from 50 000 to 120 000 ppm) was found, degradation was very slow or nil; and a combined treatment has been proposed. The superficial part to be treated by biopiles, reducing from 120 000 to 5 000 ppm in 6 months (Rojas *et al.*, 2005) and to leave the remainder for NA treatment, Active treatment of the surface layer of the deposits would accelerate the process and reduce the costs of treating this type of sites, leaving the natural process to take care of most of the volume to be treated.

Table 4. Decrease (natural attenuation) in TPHs, considered as the diminution per level during the follow-up period. Both sites were divided in four depths as a function of the encountered TPHs concentration (A, 0-60 cm; B, 60-120 cm; C, 120-180 cm; D, 180-220 cm)

	Decrease of TPHs (mg/kg of ds*/year)	
	Paredón 31	Campo 10
Level A	0	1122
Level B	726	2257
Level C	220	167
Level D	125	6

*ds = dry soil

References

[1] Aire Force Center for Enviroinmental Excellence, 1999., atural Attenuation of Fuel Hydrocarbons Performance fron Multiple Air Force Demostration Sites. Technology Demsotration Technical Summary Report.

[2] Atteia O., y C. Guillota, 2007. Factors controlling BTEX and chlorinated solvents plume length under natural attenuation conditions. *Journal of Contaminant Hydrology* 90(1-2): 81-104

[3] Bento F.M., F. A. Camargo, B. C. Okeke, and W. T. Frankenberger. 2005. Comparative bioremediation of soils contaminated with diesel oil by natural attenuation, biostimulation and bioaugmentation. Bioresource. *Technology* 96: 1049-1055

[4] Chapman S. W., B. L. Parker, J. A. Cherrya, R. Aravena, and D. Hunkeler. 2007. Groundwater–surface water interaction and its role on TCE groundwater plume attenuation. *Science of the Total Environment* 379(2-3): 133-150.

[5] Chen K.F., C.M. Kao, J.Y. Wang, T.Y. Chen and C.C. Chien. 2005. Natural attenuation of MTBE at two petroleum-hydrocarbon spill sites. *Journal of Hazardous Materials* 125, (1-3): 10-16.

[6] Davis J.W., J. M. Odom, K. A. DeWeerd, D. A. Stahl, S. S. Fishbain, R. J. Westa, G. M. Klecka and J. G. DeCarolis. 2002. Natural attenuation of chlorinated solvents at Area 6, Dover Air Force Base: characterization of microbial community structure. *Journal of Contaminant Hydrology* 57(1-2): 41-59

[7] Dupont R.R., D. L. Sorensen, M. Kemblowski, M. Bertleson, D. McGinnis, I. Kamil, and Yang Ma. 1998. EPA Project Summary Monitoring and Assessment of In-Situ Biocontainment of Petroleum Contaminated Ground-Water Plumes. EPA/600/SR-98/020.

[8] Environmental Protection Agency (EPA). 1992. United States Environmental Protection Agency. Supplemental Guidance to RAGS: Calculating the Concentration Term, Publication 9285.7-08, NTIS No. PB92-963373.

[9] Environmental Protection Agency (EPA). 1991. United States Environmental Protection Agency. Risk Assessment Guidance for Superfund: Volume I - Human Health Evaluation Manual (Part B, Development of Risk-based Preliminary Remediation Goals. EPA/540/R-92/003.

[10] Environmental Protection Agency (EPA). 1999. Use of Natural Attenuation at Superfound, RCRA Corrective Action, and Undergraund Storage Tank Sites., OSWER 9200.4-17P

[11] Eweis, J.B., S.J. Ergas, D.P. Chang,, and E.D. Schroeder. 1999. Principios de Biorremediación, Tratamiento para la descontaminación de suelos y aguas subterráneas mediante procesos biológicos y físico-químicos. McGraw Hill Interamericana. España

[12] Fernández Linares L., N.G. Rojas-Avelizapa, T.G. Roldan-Carrillo, M E. Ramírez Islas, R. J. Reyes Ávila, R. Uribe Hernández, H. G. Zegarra Martínez, J. M. Arce Ortega. 2004. Atenuación natural de suelos contaminados con hidrocarburos. Informe final (CD). Programa de Biotecnología del Petróleo, Instituto Mexicano del Petróleo.

[13] Ferrey, M., Estuesta P., Brandon B. and Nalpinski M. 1998. Measuring Mn+2 in ground water natural attenuation studies In: Godage B. Wickmanayake and Robert E. Hinchee (Eds.) Natural Attenuation. Battelle Press

[14] Gault A.G., D. R. Cooke, A. T. Townsend, J. M. Charnock and D. A. Polya. 2005 Mechanisms of arsenic attenuation in acid mine drainage from Mount Bischoff, western Tasmania. *Science of The Total Environment 345 (1-3) 219-228*

[15] Gallego J.R., E. González-Rojas, A.I. Peláez, J. Sánchez, M.J. García-Martínez, J.E. Ortiz, T. Torres and J.F. Llamas. 2006. Natural attenuation and bioremediation of Prestige fuel oil along the Atlantic coast of Galicia (Spain). *Organic Geochemistry* 37(12): 1869-1884

[16] Heller A. and J.R. Brock. 1999. Materilas and methods for photocatalyzing oxidation of organic compound in water, United States Patent, 5: 256616.

[17] IUPAC, 1997. IUPAC Compendium of Chemical Terminology mostly corresponds to the second edition. Compiled by Alan D. McNaught and Andrew Wilkinson (Royal Society of Chemistry, Cambridge, UK). http://www.iupac.org/publications/compendium/index.html.

[18] Jacqout F., M. Guiliano, P. Doumenq, D. Mundz, and G. Mili. 1999. n vitro Photooxidation of crude oil maltenic fraction: Evaluation of fósil Biomarkers and polycyclic aromatic hydrocarbns. *J. Chemosphre* 33(4): 671-681.

[19] Kennedy, Lonnie G; Jess W. Everettb, and James Gonzalesc. 2006. ssessment of biogeochemical natural attenuation and treatment of chlorinated solvents, Altus Air Force Base, Altus, Oklahoma. *Journal of Contaminant Hydrology*. 83 (3-4): 221-236

[20] Khan F.I. y T. Husain. 2001. Risk-based monitored natural attenuation a case study *Journal of Hazardous Materials* 85(3): 243-272

[21] Laane R. W., W.W. Gieskes, and G.W. Kraay. 1985. Oxygen consumption from Natural water by photo-oxidizing processes. *Neth. J. Sea Res*. 19: 125-132

[22] Lacaze J.C., and O. Villedon De Naide. 1976. Influence of llumination on phytotoxicity of crude oil. *J. Marine Pullotion Bulletin* 7: 73-77

[23] Lenczewski M., P. Jardine, L. McKay and A. Layton. 2003.Natural attenuation of trichloroethylene in fractured shale bedrock. *Journal of Contaminant Hydrology* 64(3-4): 151-168

[24] Maguire M., J. Slavek, I. Vimpany, F.R. Higginson and W.F. Pickering. 1981. Influence of pH on cooper and zinc uptakeby soil clay. Aust J. Soil Res. 19: 217-229.

[25] Member Agency of the Federal Remediation Technologies Roundtable (MAFRTR). 2004. Abstract of Remediation Case Studies, Vol. 8.

[26] Member Agency of the Federal Remediation Technologies Roundtable (MAFRTR).. 2001. Abstract of Remediation Case Studies, Vol. 5.

[27] Member Agency of the Federal Remediation Technologies Roundtable (MAFRTR).. 2000. Abstract of Remediation Case Studies, Vol. 4

[28] National Research Council. 2000. Naturla Attenuation for Groundwater Remediation, Committee on Intrinsic Remediation, Water Science and Technology Board and Board on Radioactive Waste Management,Commission on Geoscience, Envirinment and Resources, National Academy Press, Washington, D.C. 274 pp.

[29] NATO/CCMS. 2003. Pilot Study Evaluation of Demonstrated and emerging Technologies for the Treatment and Clean Upo f Contaminaed Lad an Groundwater (Phase III) 2002. Annual Report, Number 255, North Atlantic Treaty Organization, EPA-542-R-02-010.

[30] Nijenhuisa, I., M. Nikolauszb, A. Kötha, T. Felföldib, H Weissc, J. Drangmeisterd, J. Großmannd, M. Kästnerb and H: Richnow. 2007. Assessment of the natural attenuation of chlorinated ethenes in an anaerobic contaminated aquifer in the Bitterfeld/Wolfen area using stable isotope techniques, microcosm studies and molecular biomarkers. *Chemosphere* 67(2) 300-311.

[31] Peña C. E., D. E. Carter, and F. Ayala-Fierro. 2001. Toxicología Ambiental, Evaluación de Riesgo y Restauración Ambiental, Southwest Hazardous Wastes Program, A superfound Basic Research and Training Programa t the Colege of Pharmacy. The University of Arizona.

[32] Pyeong-koo L., Kang M., Choi S. H. and Touray J.C. 2005. Sulfide oxidation and the natural attenuation of arsenic and trace metals in the waste rocks of the abandoned Seobo tungsten mine, Korea. *Applied Geochemistry* 20(9): 1687-1703

[33] Qstgaard, K., A. Auborga, J. Klungsoyr and Jensen. 1987. A Comparative Studies of phytotoxicity and chemical composition of aqueous Oil affected by evaporation, illumination and extraction. *J. Water Res.* 21: 155-162.

[34] Ravi V., J.S. Chen, J. T. Wilson, J. A. Johnson, W. A. Gierke and L. A. Murdie 1998. Evaluation of Natural Attenuation of Benzene and Dichloroethanes at the KL Landfill. Bioremediation Journal 2(3): 239-258

[35] Rojas-Avelizapa N.G., T.G. Roldán-Carrillo, H.G. Zegarra-Martínez and L.C. Fernández-Linares. (2007). A field trail for an exsitu bioremediation of a drilling mud polluted site. *Chemosphere* 66: 1594-1600.

[36] Renner, R. 2000. Natural Attenuation's popularity outpace scientific sipport, NRC finds. *Environ. Sci. Tehnol.* 34: 203A-204A.

[37] Serrano A., M. Gallego, and J.L. González. 2006. Assessment of natural attenuation of volatile aromatic hydrocarbons in agricultural soil contaminated with diesel fuel. *Environmental Pollution* 144(1): 203-209.

[38] Schirmer M., B. J. Butler, J. F. Barker, C. D. Church and K. Schirmer. 1999. Evaluation of biodegradation and dispersion as natural attenuation processes of MTBE and benzene at the Borden field site. Physics and Chemistry of the Earth, Part B: Hydrology, Oceans and Atmosphere 24(6): 557-560

[39] Scow K.M., and K. A. Hicks. 2005. Natural attenuation and enhanced bioremediation of organic contaminants in groundwater. *Current Opinión in Biotechnology* 16: 246-253

[40] Snape I., S. H. Ferguson, P. McA. Harvey, and M. J. Riddle. 2006. Investigation of evaporation and biodegradation of fuel spills in Antarctica: II—Extent of natural attenuation at Casey Station. *Chemosphere* 63(1): 89-98

[41] Stollenwerk G.K., N. G. Breit, A.H. Welch, J.C Yount, J.W. Whitney, A.L. Foster, M. N. Uddin, R.K. Majumder and N. Ahmed. 2007. Arsenic attenuation by oxidized aquifer sediments in Bangladesh. *Science of The Total Environment* 379(2-3): 133-150

[42] Taghvaei Ganjali S., B. Nahri Niknaf, M. Khosravi. 2007. Photooxidation of crude petroleum maltenic fraction in natural simulatd conditionsand structural elicidation of photoproducts. Iran. *J. environ. Healh Sci. Eng.* 4(1): 37-42

[43] USEPA, 1999. Use of Monitored Natural Attenuation at Superfund, RCRA corrective action, and underground storage tank sites". OSWER Directive 9200.4-17P. USEPA

[43] U.S. Army Science Board. 1995. Infrastructure and Environmental Sigue Group, Remediation of Contaminated Army Sites: Utility of Natural Attenuation, Draft Report, Department of the Army, Washington D.C.

[44] Wang S., and C.N. Mulligan. 2006. Natural attenuation processes for remediation of arsenic contaminated soils and groundwater. *Journal of Hazardous Materials* 138(3): 459-470

[45] Yong, R.N. 2000. Geoenvironmental Engineering: Contaminated Soils, Pullutante Fate and Mitigation. CRC Press, Boca Raton, FL. 307 p

[46] Young R.N. B.P., Warketin, Y. Phadungchewit and R. Galvez. 1990. Buffer capacity and lead retention in some clay materials. *Water Air Soil Pollution* 53: 53-67.

In: Remediation of Soils and Aquifers
Editors: Luis G. Torres and Erick R. Bandala
ISBN 978-1-60741-372-1

Chapter 6

APPLICATION OF ENVIRONMENTAL RISK ASSESSMENTS IN MEXICO

Rosa Maria Flores Serrano
Instituto de Ingenieria-UNAM

1.1. Regulatory policies

There are different regulations in Mexico that have as main purpose the protection of soil, water, and other natural resources. Generally, these regulations are within the context of the management of hazardous substances and materials, and high risk activities (industries that use hazardous substances in such quantities that require some kind of control, known as "report quantities"). The most important law is the General Law of Ecological Balance and Environmental Protection (Ley General del Equilibrio Ecológico y la Protección al Ambiente, LGEEPA) published for the first time in 1988. This law has some bylaws (covering the details on how to implement the diverse aspects of the law) and the Official Mexican Standards (Normas Oficiales Mexicanas, NOMs), which are aimed at enforcing the law. NOMs in general, are technical regulations that establish the rules, procedures, specifications, limits, or any other characteristics applicable to a service, operation, installation, procedure, product or activity.

Article 136 of the LGEEPA deals with the prevention and control of soil contamination, and establishes that wastes deposited in soil must comply with the necessary conditions to avoid: i) soil contamination, ii) adverse alterations of the biological processes that take place in the soil, iii) perturbations that prevent soil from being used or exploited, and iv) risks and health problems (LGEEPA, 1998). Nevertheless, the terms risk, environmental risk, or environmental risk assessment are never defined in that legislation. Article 152 of this law indicates that "when the generation, handling, and disposal of hazardous materials and wastes result in the contamination of the soil, the persons responsible for this contamination must take the appropriate actions to recover the conditions of the soil in order to meet the requirements of the land-use or ecological management plans"; nevertheless, until the year 2002, there were no Mexican standards that could be used to determine when a soil was contaminated and the soil clean-up levels. This fact produced controversies between the authority and those responsible for the contamination. These controversies were solved on a case by case basis, sometimes using foreign standards, *e.g.*, region IX Preliminary Remediation Goals (PRGs) from the United States Environmental Protection Agency (U.S.EPA), or by performing human health risk assessments. The results were accepted as a goodwill agreement between the parts.

It is important to say, at this point, that the Environmental Impact Assessment Bylaw (EIAB) of the LGEEPA states in articles 17 and 18 that for high risk activities it is necessary to perform a risk assessment study. This risk assessment study must consider the scenarios and preventive actions resulting from the risk analysis of the project. As in LGEEPA, the terms risk, environmental risk, risk analysis, or risk assessment are not defined in this bylaw. SEMARNAT (2002) states that "environmental risk" in the EIAB means the probability that major accidents will occur in a facility due to the hazardous materials management in the high risk activities; accidents may trespass the facilities' boundaries, affecting human populations, properties, the environment, and ecosystems. The risk assessment of these accidents must consider the radius of the effects and their intensity. There are guidelines proposed by the Ministry of Environment and Natural Resources (Secretaría del Medio Ambiente y Recursos Naturales, SEMARNAT) to perform these risk studies with the inconvenient that, in the guidelines, a definition of Environmental Risk Assessment is given: "a preventive tool that promotes the systematic application of policies, procedures, and management practices to the analysis, assessment and control of risks, with the purpose of protecting the population and the society, anticipating the possibility of accidental spills of substances considered hazardous. The study evaluates the potential impact of these substances and promotes its prevention or mitigation". But clearly the environmental risk assessment addressed in the bylaw and the guidelines mentioned above is not the same risk assessment of concern for contaminated sites. This duplicity of the terminology sometimes causes confusions among users. It is expected that in a near future the Mexican Ministry of Environment will clarify this issue.

Considering these problems, the need to develop standards to establish soil cleaning levels and other regulations for the application of environmental risk assessments was obvious. In 1998, the Mexican Federal Bureau of Environmental Protection (Procuraduría Federal de Protección al Ambiente, PROFEPA) published the Interim Criteria for Polluted Soil Restoration (PROFEPA, 1998). These criteria gathered Maximum Allowable Limits (MALs) from international agencies (*e.g.*, U.S.EPA) for some hydrocarbons and inorganic substances in the soil but was not compulsory, since it was not established as a NOM; nevertheless it did help a lot in the management of contaminated sites. Finally, in the year 2002 the Emergency NOM named NOM-EM-138-ECOL-2002, establishing the MALs for pollution of soils affected by hydrocarbons, site characterization, and restoration procedures (DOF, 2002a; LEXCORP, 2008).This emergency standard considered three kind of products (gasoline, diesel, and oil products, measuring Total Petroleum Hydrocarbons, TPHs, for each kind of product), and three kinds of land uses (industrial/commercial, residential, and for agriculture/ recreational/habitat-conservation purposes). This standard compiled some of the information contained in the Interim Criteria and was in force between August 21, 2002 and August 20, 2003; this was due to the fact that Emergency Norms, following the provisions of the Law of Measurements and Standardization of Mexico, may only be in force for 6 months and extended only once for another equal term (Baker and McKenzie, 2005). The standard stated that it was possible to use MALs different from those established in this regulation only when the one responsible for the contamination proved to the authority that the proposed cleaning levels posed no risk for the human population, flora, and fauna at the site. It was not clearly said, but it could be interpreted that the application of environmental risk assessments was allowed.

In the same year, 2003, the General Law for the Prevention and Integral Management of Wastes (Ley General para la Prevención y Gestión Integral de los Residuos, LGPGIR) was published and is still in effect. This law considers the obligation to clean-up contaminated sites in order to meet the NOMs (if there is one for the substance of interest) or the concentrations that do not pose a risk to human health and the environment (risk based clean-up levels). This law defines the terms "risk" and "environmental risk assessment".

The emergency standard NOM-EM-138-ECOL-2002 was substituted in the year 2005 for the Official Mexican Standard, NOM-138-SEMARNAT/SS-2003, entitled "Maximum allowable limits of hydrocarbons in soil and specifications for their characterization and remediation" (DOF, 2005), and it remains in force. This new standard differs from the former emergency one in some aspects. It is less stringent with regard to maximum allowable limits for certain substances, and establishes new methods of analysis (now light, medium, and heavy hydrocarbon fractions are measured instead of TPHs). It also contemplates some restrictions with regard to remediation (i.e., remediation by genetically modified microorganisms that were absent in the Emergency Standard) (Baker and McKenzie, 2005). It allows the development of environmental risk assessments in order to establish site-specific cleaning levels, as well as the reduction of the bioavailability of the substances or their exposure; this applies in the case of environmental passives and other cases that must be justified to the authority. In the year 2007, the Official Mexican Standard NOM-147-SEMARNAT/SSA1-2004 entitled "Criteria for establishing the concentrations of soil remediation polluted by arsenic, barium, beryllium, cadmium, chrome, mercury, nickel, silver, lead, selenium, thallium and/or vanadium" was published (DOF, 2007). After almost ten years since the publication of the Interim Criteria, the problem of metals was again considered in a regulation. It is important to mention that the government had been working on this standard since the year 1998, but changes in the administration and other technical issues made its publication impossible before. This standard contemplates four kinds of remediation goals: MALs established by the authority, background concentrations, site-specific remediation goals based on environmental risk assessments, and soluble fractions established by the authority.

With the enactment of NOM-138 and NOM-147, the provisions of the LGPGIR regarding liability for pollution and remediation may currently be implemented properly, at least with regard to soil pollution caused by some hydrocarbons and the twelve metals specified in both standards. Regarding groundwater, no NOMs to regulate the MALs of contaminants exist yet. There is only one guideline entitled "Ecological Criteria for Water Quality" (DOF, 1989); this guideline establishes MALs for different water uses, including drinking water sources, and for different contaminants (some volatile organic compounds, metals, and pesticides). The problem is the same that prevailed for the Interim Criteria for Polluted Soil Restoration: these criteria are not compulsory, since they are not a NOM. In cases where groundwater is compromised, an environmental risk assessment is a good tool to say how clean is clean.

In the year 2006, SEMARNAT published the bylaw supporting the LGPGIR (DOF, 2006), giving some details about the information that must be addressed in the environmental risk assessments; in the same year, the "Technical guideline to orientate the elaboration of environmental risk assessments for contaminated sites" was published (Technical Guideline for ERA) (SEMARNAT, 2006). With these two instruments, the Mexican government reinforces the application of environmental risk assessments, both legally and technically.

1.2. Environmental Risk Assessment Concepts

Even though risk refers to probability (the probability that a harm will occur); in an environmental context, risks are not always expressed in these terms, *e.g.*, for non-carcinogenous substances the risk estimate is a quotient and not a probability. Considering this fact, the Mexican legislation in the LGPGIR defines *environmental risk assessment* (ERA) as "the methodology to determine the probability or possibility that an adverse effect will occur to the living organisms as a consequence of their exposure to the substances or infectious agents contained in the hazardous wastes" (DOF, 2003).

There are three issues about this definition that are important to highlight: a) the inclusion of the word possibility; b) the use of the term "living organisms" to define the target receptors; and c) it is directed to the hazardous wastes. The word possibility permits to establish risk estimates not only as a probability but also in terms of quotients or qualitative descriptions that otherwise would not be accepted by the authority (sometimes when natural resources are assessed it is not always possible to present numerical values but a good description is better than no assessment). In some parts of the world, *e.g.*, USA, two kinds of assessments are distinguished depending on the type of receptor: Human Health Risk Assessments for human individuals or populations, and Ecological Risk Assessments for the study of risks to the natural environment (Barnthouse and Suter, 1986). Considering "the living organisms" as the target receptors, the environmental risk assessment, as defined above, establishes both kinds of assessments in one instrument. Finally, the fact that the assessment is considered within the context of hazardous wastes denotes that the main objective of these assessments is the protection of potential receptors against the effect of the toxic substances. Both, the bylaw supporting the LGPGIR and the Technical Guideline for ERA, support this definition.

The Technical Guideline for ERA is not mandatory, as its name says it is only a guideline, but it is a good tool for the risk assessors to accomplish what the law says; the assessors may promote different approaches as long as they meet the guideline criteria

The Technical Guideline for ERA establishes four levels of assessment, as shown on Table 1.

Levels 1, 2, and 3 are based on the hazard quotient (HQ) method that relates an exposure concentration or dose to a Toxicity Reference Value (TRV) considered to represent a "safe" environmental value (concentration or dose). HQs ≤ 1.0 indicate acceptable risks, whereas HQs > 1.0 indicate unacceptable risks. Levels 1 and 2 are deterministic and level 3 may have a probabilistic approach. Level 0 means that there is no need of ecotoxicological information, since there are no wildlife or other flora and fauna elements to be protected (even though the soil contains invertebrates, at this level they are not considered). Level 1 indicates that some TRVs are needed, but they are not available for the substance and the organism to be assessed. In this case, it is recommended to use surrogate values and the application of extrapolation equations describing the uncertainties associated to this procedure. Level 2 applies when ecotoxicological information is available for the substance to be assessed or the organism of interest; the application of extrapolation equations may be needed as well but compared to level 1, it is expected that the uncertainties may be lessened; it can also be chosen to develop ecotoxicological essays in order to obtain specific TRVs (in this case level 2 ERA turns into level 3). Level 3 applies when HQ > 1.0 was obtained after performing a level 2 ERA. For levels 1 and 2 the endpoints are related to the organism level whereas for level 3 effects at higher levels of organization are looked for. The use of exposure and effect biomarkers is encouraged in order to gain a better understanding of the cause-effect relationship.

Table 1. Levels of assessment in the Technical Guideline for ERA

Type of receptor	Type of area		
	Urban (Industrial/ Commercial)	Not-urban (Agriculture)	**Protected natural area (valuable ecosystems)
Non-biological natural resources (soil, water)	Level 0 Case A	Level 0 Case A	Level 0 Case A
Non-biological natural resources and human populations	Level 0 Cases A and B	Level 0 Cases A and B	Level 0 Cases A and B
Natural resources (biological and non-biological). Species with economic value. Without toxicological information.		Level 1 Consider Case B	Level 1 Consider Case B
Natural resources (biological and non-biological). *Threatened/Protected species or with an economical value. With toxicological information.		Level 2 Consider Case B	Level 2 Consider Case B
Natural resources (biological and non-biological). *Protected species and ecosystems.			Level 3 Consider case B

Source: Adapted from SEMARNAT (2006)

*NOM-059-ECOL-2001 (DOF, 2002b) gives a list of the Mexican species (flora and fauna) that are the object of some "risk" or "protection" category (threatened, in danger of extinction, special protection status, probably extinguished from wild environments).

** In Mexico there are six categories of Protected Natural Areas (National Parks, Biosphere Reserves, Protected Flora & Fauna Areas, Nature Sanctuaries, Protected Natural Resource Areas, and Natural Monuments); these areas are protected and managed by the National Commission of Protected Natural Areas (Comisión Nacional de Áreas Naturales Protegidas, CONANP), a federal agency.

Regarding terrestrial plants, the guideline considers the analysis of contaminants in the edible parts of the vegetation (if they are considered within the food chain); on the other hand, application of biotic indexes (*e.g.*, richness, biodiversity, relative importance value) is considered, aimed at comparing two ecosystems (*e.g.*, the impacted system of interest *vs*. a non-impacted system of similar characteristics) or to determine changes over time in a given system (*e.g.*, the impacted system before and after the stressor agent appeared).

Case A and B in the guideline are used to designate the information needed to perform the ERA when there are no human or wildlife populations to be considered (case A) and when a human population is assessed (case B).

This framework is of recent application and, in the next section, a case study where this guideline was applied is briefly presented. The project addressed both human and ecological receptors but the results presented are only for risks to the natural environment. The experience was useful to identify research lines to be reinforced in this field in Mexico.

1.3. Case Study

The site is located in the northeast part of Mexico. Details of the location are not given due to confidentiality of the information. At the beginning of the XX century, there was an accident related to an oil well. Thousands of barrels of crude were spilled. At the present time, most of the product is highly weathered and, in part, has been covered by natural vegetation (355 Ha). The spill also affected a pond that is located to the east, approximately at 2 km from the well, but in this study only the terrestrial environment was considered. The inland affected zone has an extension of 867 Ha. Most of the land has no use (745 Ha that are not used for residential, agricultural or commercial purposes) because it becomes flooded during the rainy season (approximately 7 months of the year) or because it is covered by natural vegetation, and some parts of the land are under both conditions. Mangrove forest, holm oak wood, coastal sand dune, and tropical semi-evergreen forest were the main types of vegetation identified at the site. The land is low, with altitudes between 4 and 9 meters above sea level (some parts are even below the sea level). Grass crops for cattle exist in the higher regions of the site (122 Ha). The stratigraphy showed that almost all the site (715 ha) is located over a superficial layer of clayey texture (1-2 m) followed by a sandy clay sub-superficial layer (> 30 m thick). There is no usable aquifer in this area of the site. The east part of the site (152 Ha) is adjacent to the pond and is located over a superficial layer of sandy clay loam (0.30 – 0.40 m) followed by a clay layer (0.40 m), a sandy clay loam layer (3.00 m), a clayey sand layer (2.00-11.00 m), and a sandy clay layer (> 30 m thick). The phreatic level in this area is located between 1.2 and 3.0 m underground. The type of climate is warm and semi-humid with rains in the summer. Annual mean precipitation and temperature are 1348.6 mm and 23.5 °C, respectively.

The main objective of the study was to perform an ERA, using the HQ method, to evaluate the risks from exposure to PAHs (polycyclic aromatic hydrocarbons) through the oral route of the wildlife present at the site. Ingestion of food, drinking water, and incidental soil was considered. Inhalation and dermal routes were not addressed because it was considered that most of the exposure would be due through the oral route. According to U.S.EPA (2005) some studies have shown that at equal exposure concentrations, inhalation of contaminants associated with dust particles is expected to contribute less than 0.1 % of the total risk as compared to oral exposure. Regarding dermal exposure, for most contaminants this route of exposure is expected to contribute less than 1 to 11 % of the total risk as compared to oral exposure.

The secondary objectives were: a) to perform an analysis of the distribution of LHF (Light Hydrocarbon Fraction), MFH (Medium Hydrocarbon Fraction), and HHF (Heavy Hydrocarbon Fraction) at the site; and b) to perform an analysis of the state of the vegetation at the site.

The HQ method was applied, and the concepts established in the Technical Guideline for ERA published by the Mexican government (SEMARNAT, 2006) were considered, as well as the basic steps to perform ERAs established by the international literature (U.S.EPA, 1998; TNRCC, 2000). In the following sections the actions taken are described. It is important to say that the results showed that more research is needed; hence, this study is considered a preliminary stage.

1.3.1. Problem Formulation and Methodology

A level 2 ERA was proposed for this project, since ecotoxicological information for some of the substances to be assessed was available.

In this stage of the study, information about the site was collected:

- Site's general information
    - Name of the site
    - Location
    - Delimitation of the area to be studied and the contaminated area
    - Type of site and land use (industrial, commercial, agricultural, for disposal of hazard or municipal wastes, for the extraction, transformation or management of oil products, mining, protected natural area, other uses)
    - Environmental components involved, including type of habitat or ecosystem (e.g. wetland, wood, pond, etc.), type of environment (terrestrial, aquatic), and type of media (soil, water, air).
    - Physiography of the site (topography, edaphology, geology, stratigraphy, physical and chemical soil properties, hydrology, climatology)
- Historical data
    - Source of contamination
    - Contaminant evolution
    - Relevant events after the contaminant event happened
    - Human activities developed at the site
    - Exposure scenarios
    - Actions taken to remediate the site
    - Present and past owners of the site

- Commercial value of the site

Most of this information is summarized in paragraphs above. Besides this information, the following information was generated:

- Site characterization
  - Soil and water (superficial) sampling and analysis
- Biological characterization
  - Type of habitats
  - Flora and fauna description
  - Inventory of the flora and fauna species at the site.
- Mobility assessment
  - Evaluation of the characteristics of the site that might result in the migration of the contaminants, *e.g.*, type of product, soil texture, porosity, geohydrological data, precipitation rate

For site characterization, a sampling plan was made. For soil, a grid of 200 m x 200 m was used. Each corner of the grid was recorded using a GPS (Global Positioning System, Garmin Etrex model). A total of 176 sampling points were obtained and soil samples were collected at three depths (0.20, 0.40, and 0.80 m). Samples were analyzed for LHF, MFH, HHF, PAHs, and BTEX (benzene, toluene, ethylbenzene, and xylenes), as defined by the standard NOM-138-SEMARNAT/SS-2003. In the case of PAHs, NOM-138 only considers six of these compounds, but, for this study, we included the 16 priority PAHs established by the U.S.EPA (acenaphthene, phenathrene, fluoranthene, fluorene, anthracene, indene(1,2,3-cd)pyrene, acenaphthylene, naphthalene, pyrene, benzo(a)anthracene, benzo(a)pyrene, benzo(a)fluoranthene, benzo(g,h,i)perylene, benzo(k)fluoranthene, chrysene and dibenzo(a,h)anthracene). For the analysis, the following methods were used: EPA-8270B-1994 for PAHs, EPA-8260B-1996 for BTEX, EPA-8015B-1996 for LHF and MHF, and EPA-1664-A-1996 for HHF. All samples were preserved as indicated in the standard NOM-138.

Water samples were collected from puddles and little ponds located within the study area. Ten water samples were collected. PAHs were extracted *in situ* using 100 ml of high purity hexane. Extracts were preserved at 4 °C and analyzed by column chromatography (gas chromatography, flame ionization detection for quantitation).

The contaminants of potential concern (COPCs) to be assessed were selected in this step of the process; ecological screening benchmarks (ESB) for soil were used for this purpose. Benchmarks provide a simple approach to compare COPCs concentrations in media at the affected property to values that represent media concentrations that are presumed to be safe to biota with the likelihood of being the most exposed. If soil concentrations were less than the benchmarks, the COPCs were excluded from further consideration, unless the octanol-water partition coefficient (Kow) showed that COPCs were potentially bioaccumulative ($6.3x10^3 <$ Kow $<1x10^8$) (TNRCC, 2001). A geometric mean of the available benchmarks for each chemical was obtained and was compared to the geometric mean of the concentrations found in the soil. Table 2 presents these values as well as the Kow coefficients.

Table 2 shows that all PAHs analyzed in soil were below the analytical detection limit (DL) established by the laboratory that performed the analysis (1 mg/kg), but they were retained for assessment since the detection limit was too high (there are ecological screening benchmarks below this limit), and they are potentially bioaccumulative. Soil concentrations considered for the assessment were the ones obtained from the upper layer because it is more likely for wildlife to come in contact with this soil. Regarding water samples, all concentrations were below 0.27 mg/l. Benzo(a)anthracene was the PAH with the highest concentration, all other PAHs presented concentrations between 0.0005 mg/l (DL/2) and 0.07 mg/l.

It is important to say that even though LHF, MHF, and HHF were retained for further evaluation, it was not possible to obtain HQ values since no TRVs were found in the literature for these hydrocarbon fractions, as defined in the Mexican standard NOM-138-SEMARNAT/SS-2003. Further studies are proposed to develop reference values for Mexican oil mixtures.

Table 2. COPCs selected for assessment based on Ecological Screening Benchmarks and its potential of bioaccumulation ($6.3x10^3 <$ Kow $<1x10^8$)

Chemical	ESB Geometric mean (mg/kg)	Kow (dimensionless)	Cs Geometric mean (mg/kg)		Status
Acenaphtene	36.02	9.22E+03	0.50	*	Retained for assessment because it is potentially bioaccumulative
Acenaphtylene	682.00	5.25E+03	0.50	*	Eliminated from further ecological review.
Anthracene	12.17	2.95E+04	0.50	*	Retained for assessment because it is potentially bioaccumulative
Benzo(a) anthracene	1.73	4.77E+05	0.50	*	Retained for assessment because it is potentially bioaccumulative
Benzo(a) pyrene	0.35	1.35E+06	0.50	*	Retained for assessment because it is potentially bioaccumulative and Cs>ESB
Benzo(b) fluoranthene	2.82	1.59E+06	0.50	*	Retained for assessment because is potentially bioaccumulative
Benzo(g,h,i) perylene	3.24	3.08E+06	0.50	*	Retained for assessment because is potentially bioaccumulative
Benzo(k) fluoranthene	148.00	1.56E+06	0.50	*	Retained for assessment because it is potentially bioaccumulative
Chrysene	0.52	5.48E+05	0.50	*	Retained for assessment because it is potentially bioaccumulative
Dibenzo(a,h) anthracene	2.23	3.53E+06	0.50	*	Retained for assessment because it is potentially bioaccumulative
Phenentrene	10.34	3.55E+04	0.50	*	Retained for assessment because it is potentially bioaccumulative

Table 2. Continued.

Chemical	ESB Geometric mean (mg/kg)	Kow (dimensionless)	Cs Geometric mean (mg/kg)	Status
Fluoranthene	3.49	1.21E+05	0.50 *	Retained for assessment because it is potentially bioaccumulative
Fluorene	39.72	1.47E+04	0.50 *	Retained for assessment because it is potentially bioaccumulative
Indene(1,2,3,c,d) pyrene	1.90	8.22E+06	0.50 *	Retained for assessment because it is potentially bioaccumulative
Naphthalene	1.82	2.36E+03	0.50 *	Eliminated from further ecological review.
Pyrene	16.74	1.00E+05	0.50 *	Retained for assessment because it is potentially bioaccumulative
Benzene	2.52	1.37E+02	0.10	Eliminated from further ecological review.
Ethylbenzene	96.24	1.33E+03	0.11	Eliminated from further ecological review.
Toluene	26.62	4.65E+02	0.11	Eliminated from further ecological review.
Xylenes	10.00	1.59E+03	0.34	Eliminated from further ecological review.
LHF defined in NOM-138	NF	5.12E+03**	6. 29	Retained for assessment because no ESB values were found for this fraction
MHF defined in NOM-138	NF	2.57E+04**	21.88	Retained for assessment because no ESB values were found for this fraction and it is potentially bioaccumulative
HHF defined in NOM-138	NF	1.66E+05**	1795.65	Retained for assessment because no ESB values were found for this fraction and it is potentially bioaccumulative

ESB = Ecological Screening Benchmark (geometric mean of diverse values found in the literature for organisms in close contact with soil, *e.g.,* invertebrates, plants, and microorganisms).

Kow = Octanol-water partition coefficient, values obtained from U.S.EPA (1999) unless something else is specified

Cs = Geometric mean of concentrations found in soil samples of the upper layer (0.20 m).

NF = Not found.

* Below analytical detection level (DL) = DL/2.

** Obtained from the equation by Lyman (1990) indicated in TPHCWG (1997): log Koc = log Kow - 0.21; log Koc values were taken from TPHCWG (1997) for the aromatic fractions >C10-C12, >C16-C21, >C21-C35 (3.5, 4.2, and 5.1, respectively).

COPCs to be assessed are marked in bold letters

For the biological characterization three field visits were conducted between the months of October and November. The duration of each field visit was 15 days. In order to obtain a list of faunal species, diurnal and nocturnal visual encounter surveys were performed all over the site. Other direct and indirect methods of identification were also used: sign surveys (*e.g.*, identification of tracks, scats, carcasses, scratches, burrows), and trapping. Tomahawk and Sherman traps were used for the collection of mammals, mist nests were used for birds and bats. Reptiles and amphibians were collected by hand or using pitfall traps, nets, and snake hooks. For invertebrates, light traps were used, as well as sweep netting and grubbing surveys.

In the case of birds and mammals, all collected specimens were released after identification, this was because in this preliminary study it was decided to use indirect methods for the estimation of contaminant concentrations in the tissues of these groups of animals, so the specimens were not killed. In the case of reptiles, amphibians, invertebrates, and vegetation, specimens were collected for the analysis of PAHs in their tissues (10 samples of each group). For this purpose six "working stations" were delimited within the study area. The stations covered a surface of 255 Ha approximately, they were located in the areas that showed the highest LHF (the most soluble fraction) concentrations; water samples were also collected in these areas. Samples were lyophilized in the laboratory. PAHs were extracted with high purity hexane and dichloromethane. Extracts were fractionated and analyzed by column chromatography (gas chromatography, flame ionization detection for quantitation).

During the field visits, 7 species of amphibians, 16 of reptiles, 18 of mammals, and 108 of birds were registered. The list of species is not shown because of space limitations. Nine species were chosen as assessment endpoints: 5 avian species and 4 mammals. Two herpetofauna species were chosen also but could not be evaluated because no TRVs were found for PAHs regarding these taxonomic groups. The criteria for the selection of these species included ecological, economical and social relevance, sensitivity, protection status under NOM-059-ECOL-2001 Mexican standard, foraging habits, and home-range size of the species. Avian species chosen were: *Buteo magnirostris, Geranospiza caerulescens, Falco femoralis, Columbina passerina,* and *Crotophaga sulcirostris*. Mammals selected for the assessment were: *Procyon lotor, Canis latrans, Sylvilagus floridanus*, and *Odocoileus virginianus*.

The flora inventory and the cartography of vegetation at a medium scale were obtained. The types of vegetation were also described. In order to accomplish these objectives foot surveys using line-transects were carried out; vehicle drives were also useful for this task. A total of 440 species were registered. All positions were recorded using a GPS system. Plots were used to obtain the Shannon-Wiener Diversity Index (DI) and the Relative Importance Value (RIV).

A conceptual model of the site was developed. The conceptual model included the following elements (the graphic representation of this model is not presented because of space limitations):

a) Source of contamination: weathered hydrocarbons from the spill.
b) Affected media: superficial soil, water from ponds and puddles, and air (the latter was not considered in the assessment, because it was expected to have a low contribution to total risk).
c) Receptors: omnivorous birds, carnivorous birds, insectivorous birds, herbivorous birds, omnivorous mammals, carnivorous mammals, insectivorous mammals, herbivorous mammals, insectivorous reptiles, herbivorous reptiles, insectivorous amphibians, terrestrial invertebrates, terrestrial plants (vegetal matter).
d) Food chain transfer: foraging habits and diet composition obtained from the literature were taken into account.

1.3.2. Analysis

The hazard quotient (HQ) for each wildlife species and COPC assessed was calculated using a dose, at which organisms at the site were exposed to, divided by a specific TRV (NOAEL, No Observed Adverse Effect Level):

$$HQ_{oral} = \frac{Dose_{oral}}{TRV_{oral}} \quad (1)$$

The dose of exposure for the oral route was calculated with the equation proposed by TNRCC (2000):

$$Dose_{oral} = \frac{\left[(IR_f \times C_f \times EMF_f) + (IR_w \times C_w \times EMF_w) + (IR_s \times C_s \times EMF_s)\right]}{BW} \quad (2)$$

Where:

$Dose_{oral}$ = estimated dose from ingestion (mg COPC/kg body weight/day)
IR_f = ingestion rate of food (prey) (kg/day)
C_f = COPC concentration in food (mg/kg)
EMF_f = exposure modifying factor for food (dimensionless)
IR_w = ingestion rate of water (L/day)
C_w = COPC concentration in water (mg/L)
EMF_w = exposure modifying factor for water (dimensionless)
IR_s = ingestion rate of soil (kg/day)
C_s = COPC concentration in soil (mg/kg)
EMF_s = exposure modifying factor for soil (dimensionless)
BW = body weight of receptor (kg)

For omnivorous species, the fraction of each type of food was considered. A modifying factor of 1 was used in all cases. Information regarding food habits, diet composition, body weights, home ranges, water content of diet items, soil, water, and food ingestion rates of the species to be assessed was gathered from the literature. Contaminant concentrations of the probable food items (prey animals and plants) were also obtained. As said earlier in this report, samples of reptiles, amphibians, invertebrates, and plants (leaves and stems) were analyzed. In the case of prey mammals and birds, the concentrations of PAHs in muscle tissue were estimated using the biotransfer factor (BTF, d/kg) for organic chemicals in beef proposed by Travis and Arms (1988). A BTF for each PAH considered in the assessment was calculated. These factors were converted to BTFs for the species of interest by dividing the BTF derived from the Travis and Arms (1988) model by the fat content of a cow and multiplying the result by the fat content of the species of interest (RAIS, 2008). The same approach was applied for birds, as proposed by U.S.EPA (1998b). Fat contents for the wildlife species of interest were not available so fat contents of surrogate species were used.

Once the BTFs for wildlife species were calculated, the BTF (d/kg) was multiplied by the species-specific food ingestion rate (kg/d) and by contaminant concentration in food (mg/kg) to obtain an estimate of the concentration in meat in milligram of contaminant/kilogram of meat. Uptake by ingestion of water and soil was also considered. The uncertainties of these estimations were high, but they were adjusted using a correction factor derived from Flores-Serrano (2005). This author measured some PAH concentrations in tissues of birds and mammals at another site with similar characteristics to the one addressed herein and then compared these results with the ones estimated with the approach described above. The results showed that the BTF underestimated PAH concentrations in mammals and birds, thus a correction factor may be calculated in order to compensate for this difference.

Food ingestion rates were estimated using the body weight-based allometric regression equations proposed by Nagy (2001). Water ingestion rates were derived from the equations proposed by Calder and Braun (1983), and soil ingestion rates were obtained from Gibson *et al.* (1997). Body weights of the species were obtained from the zoological collections of the National Autonomous University of Mexico (Universidad Nacional Autónoma de México, UNAM), and the National Polytechnical Institute (Instituto Politécnico Nacional, IPN).

TRVs were derived from laboratory tests found in the literature. In all cases it was necessary to extrapolate the toxicity data from the laboratory test species to the wildlife species. This was achieved by using the allometric scaling approaches proposed by Sample *et al.* (1996), and Sample and Arenal (1999). Reproductive and developmental endpoints were preferred because they may have a greater impact on a population or community scale.

1.3.3. Results (Risk Characterization) and Discussion

With the exposure doses estimated and the TRVs derived for each endpoint species, the HQs were calculated as shown on equations 1 and 2. All HQs were below 1, so it was considered that the risks from exposure to PAHs via oral route were acceptable. LHF, MHF, and HHF were not assessed because of the lack of toxicological information for these groups of substances. Further investigation is needed to develop appropriate TRVs and ESBs for these hydrocarbon fractions as defined in the standard NOM-138.

Since most of the hydrocarbons found on the site were strongly weathered and the number of wildlife species found in the short time of field work was not very different from the ones found at sites not affected by hydrocarbons (see Bojorges *et al.*, 2006), alternative *in situ* ecotoxicological studies are proposed. These *in situ* studies (*e.g.*, biomarkers effect, reproductive success, rodent sperm analysis), would allow establishing the likelihood that toxic effects that may compromise the survival of the species are indeed occurring, because some investigations suggest that the bioavailability and toxicity of chemicals in soil becomes greatly compromised with time (Kaputska, 2004; Tannenbaum, 2003).

In the case of vegetation, the Relative Importance Value (RIV) showed that *Avicennia germinans* was the most important mangrove species in the ecosystem, but *Conocarpus erectus* was the most important species in the areas where LHF and HHF presented the highest concentrations. It was observed that mangrove repopulation is taking place even in the affected areas with high concentrations of hydrocarbons. However, comparing a mangrove forest area with very low LHF and HHF concentrations to the one that presented the highest concentrations, the Shannon-Wiener DI showed that the latter had the highest DI; an opposite behavior was expected, this difference was attributed to the fact that the more contaminated site was the most proximate to the tropical semi-evergreen and the holm oak wood forests.

Sites with intermediate concentrations presented the lowest DI values. Density and height of the specimens were also measured on the mangrove forest vegetation. Low densities and heights were observed in the areas with higher hydrocarbon concentrations.

These results suggest that the presence of hydrocarbons may have an adverse impact on diversity, density, and structure (height) of the vegetation, but there might be other environmental agents that could have the same impact as well (*e.g.*, salinity, surrounding vegetation, and nutrients). Further investigation is needed to quantify the contribution of hydrocarbon concentrations and other environmental factors on the vegetation structure.

An assessment of the pond was also recommended to determine the potential migration of contaminants from the inland to the pond, and the effects on the aquatic organisms, including the aquatic avifauna.

1.4. Other Mexican Experiences

Since the 1980s, many efforts have been made in Mexico to obtain ecotoxicological information that may help to establish environmental benchmarks, most of them are addressed to the aquatic environment and terrestrial earthworms. Some of the issues considered in the investigations, among others, are: estimation of acute lethal concentrations, measurement of PAH metabolites, AChE (acetylcholine esterase) and BChE (butyrylcholine esterase) activity, cytochrome P-450, and EROD activity, the formation of DNA adducts, DNA damage (comet assays), and MT (metallothioneins) induction (see the works of Vázquez *et al.*, 1986; Martínez-Tabche *et al.*, 1999; Pica-Granados *et al.*, 2000; Zapata-Pérez *et al.*, 2000; Galindo-Reyes *et al.*, 2002; Gold-Bouchot *et al.*, 2006; Hernández-Zamora *et al.*, 2006; Santos-Medrano *et al.*, 2007).

Lately, more efforts have been made to perform studies addressing assessments of the adverse effects on terrestrial vertebrates, *e.g.*, the works performed by Flores-Serrano, 2005; Jiménez *et al.*, 2005; García-Hernández *et al.*, 2006; Jasso-Pineda *et al.,* 2007. The work by Flores-Serrano (2005) was similar to the case presented in the present document, where the HQ method was applied to assess risks from exposure to PAHs through the oral route for the birds and mammals. The investigation of Jasso-Pineda *et al.*, 2007, integrated an environmental assessment together with the study of the exposure to heavy metals from mining sites and the analysis of DNA damage (comet assay) in both children and local fauna (rodents). Their results showed a correlation with increased DNA damage in the most exposed animals. The works by Jiménez *et al.* (2005) and García-Hernández *et al.* (2006) are focused on the accumulation of metals and organochlorine contaminants in birds' liver and AChE activity in birds, respectively.

Nevertheless these investigations are mainly for metals and pesticides, but there are not many for hydrocarbons. It is also remarkable that it is very difficult to find investigations where the effect (potential or actual) is related to a safe concentration in the natural environment in order to protect populations or communities. Although these studies give an idea about the health of organisms in the system by measuring both effect and exposure biomarkers or concentrations of contaminants in the tissues, they do not mention anything about the survivability of the studied populations.

That is why it is important to say that biomarkers should not be used as a replacement for conventional monitoring techniques, but should be applied as supplementary approaches for demonstrating links between sublethal biochemical concentrations and adverse effects in natural populations in field studies. The research challenge for using biomarker measurements is to predict effects at the population level from effects at the individual level measured upon individuals collected in the field (Hyne and Maher, 2003).

Long-term monitoring programs of wildlife populations, inventories of the biodiversity, information about endangered species, data about body weights, diet compositions, foraging habits, among other ecological information, are also necessary to obtain more accurate risk estimates and know the real impact of the substances. In Mexico, there have been few efforts to integrate this kind of information at a national level. The National Commission for the Knowledge and Management of Biodiversity (CONABIO, Comisión Nacional para el Conocimiento y Uso de la Biodiversidad, www.conabio.gob.mx) and SEMARNAT, through the National System of Environmental Information and Natural Resources (Sistema Nacional de Información Ambiental y de Recursos Naturales, SNIARN, http://www.semarnat.gob.mx/informacionambiental/Pages/index-sniarn.aspx), has useful data bases but they are not enough to fulfil the needs of the risk assessors. Universities are other important source of information but sometimes it is difficult to get to know what they are working on, and to have access to their research results. It is expected that in a near future the results obtained by the environmental risk assessors and ecotoxicologists may help to improve the data bases.

There are also other forums where it is possible to have a taste of the advances in the field of the environmental risk assessment in Mexico. The Mesoamerican Association of Ecotoxicology and Environmental Chemistry (AMEQA, Asociación Mesoamericana de Ecotoxicología y Química Ambiental, http://www.ameqa.org/), the Mexican Federation of Sanitary Engineering and Environmental Sciences (FEMISCA, Federación Mexicana de Ingeniería Sanitaria y Ciencias Ambientales A.C., http://www.femisca.org/), the Latin American Network for the Prevention and Management of Contaminated Sites-Chapter Mexico (RELASC-Mexico, Red Latinoamericana de Prevención y Gestión de Sitios Contaminados, http://relascmex.wordpress.com/) of recent creation, are some of the organizations that promote periodical meetings to disseminate the research in different disciplines that have something to do with the environment, including risk assessment.

In summary, there is still a lot of work to be done in the field of environmental risk assessment, but we are constantly improving, both legally and technically.

References

[1] Baker and McKenzie Abogados S.C. (2005), "NOM 138: Soil polluted by hydrocarbons/NOM 138: Suelos contaminados por hidrocarburos", Environmental and Natural Resources, April 2005, Number 2, URL: *http://www.bakernet.com/NR/rdonlyres/1157C227-F0DD-45D1-89AB-309AD2440BA6/38295/02NOM138.pdf* (consulted on June 2008).

[2] Barnthouse, L W y Suter II, G W (1986) (eds), *User's manual for ecological risk assessment*, ORNL-625, Oak Ridge National Laboratory, Oak Ridge.

[3] Bojorges, J B, López-Mata, L, Tarango-Arámbula, L A, Herrera-Haro, J G y Mendoza-Martínez, G D (2006), "Combinación de métodos de muestreo para registrar la riqueza

de especies de aves en ecosistemas tropicales", *Universidad y Ciencia Trópico Húmedo*, v 22, No 2, 111-118.

[4] Calder, W A and Braun, E J (1983), "Scaling of osmotic regulation in mammals and birds", *Am. J. Physiol.*, V 244, R601-R606.

[5] DOF (Diario Oficial de la Federación) (1989), "Criterios ecológicos de calidad del agua", *Diario Oficial de la Federación*, 13 de diciembre de 1989, México D.F.

[6] DOF (Diario Oficial de la Federación) (2002a), Norma Oficial Mexicana de Emergencia NOM-EM-138-ECOL-2002, que establece los límites máximos permisibles de contaminación en suelos afectados por hidrocarburos, la caracterización del sitio y procedimientos para la restauración", *Diario Oficial de la Federación*, 20 de agosto de 2002, México D.F

[7] DOF (Diario Oficial de la Federación) (2002b), *Norma Oficial Mexicana NOM-059-ECOL-2001 Protección Ambiental-Especies Nativas de México de Flora y Fauna Silvestres–Categorías de Riesgo y Especificaciones Para su Inclusión, Exclusión o Cambio-Lista de Especies en Riesgo*, DOF del 6 de marzo de 2002, México D.F.

[8] DOF (Diario Oficial de la Federación) (2003), "Ley general para la prevención y gestión integral de los residuos ", *Diario Oficial de la Federación*, 8 de octubre de 2003, México D.F.

[9] DOF (Diario Oficial de la Federación) (2005), "Norma Oficial Mexicana NOM-138-SEMARNAT/SS-2003, límites máximos permisibles de hidrocarburos en suelos y las especificaciones para su caracterización y remediación", *Diario Oficial de la Federación*, 29 de marzo de 2005, México D.F.

[10] DOF (Diario Oficial de la Federación) (2006), Reglamento de la ley general para la prevención y gestión integral de los residuos, *Diario Oficial de la Federación*, 30 de noviembre de 2006, México D.F

[11] DOF (Diario Oficial de la Federación) (2007), "Norma Oficial Mexicana NOM-147-SEMARNAT/SSA1-2004, que establece criterios para determinar las concentraciones de remediación de suelos contaminados por arsénico, bario, berilio, cadmio, cromo hexavalente, mercurio, níquel, plata, plomo, selenio, talio y/o vanadio", *Diario Oficial de la Federación*, 2 de marzo de 2007, México D.F.

[12] Flores-Serrano, R M (2005), *Evaluación de riesgo ecológico en un sitio contaminado por hidrocarburos en México*, Ph.D. Thesis, Programa de Maestría y Doctorado en Ingeniería, Universidad Nacional Autónoma de México, México D.F.

[13] Galindo-Reyes, J G,. Leyva, N R, Millan, O A, and Lazcano, G A (2002), "Effects of pesticides on DNA and protein of shrimp larvae Litopenaeus stylirostris of the California Gulf", *Ecotoxicology and Environmental Safety*, v 53, 191-195.

[14] García-Hernández, J, Cadena-Cárdenas, L, Villegas-Álvarez, M M, Moreno-Ramos, T M, Holguín-Acosta, E (2006), "Exposición de aves migratorias neotropicales de sonora y sinaloa a plaguicidas organofosforados", *Memorias del 2° Congreso de la Asociación Mesoamericana de Ecotoxicología y Química Ambiental (AMEQA),* 24-28 de Abril, 2006, Benemérita Universidad Autónoma de Puebla, Puebla, 88.

[15] Gibson, E, Strudwick, D and Walker, Ph (1997), *National framework for ecological risk assessment of contamminated sites: Part B: derivation ecological impact levels (EILs)*, Victorian Environmental Protection Authority, Environment Australia, Australia.

[16] Gold-Bouchot, G, Zapata-Perez, O, Rodriguez-Fuentes, G, Ceja-Moreno, V, Del Rio-Garcia, M, and Chan-Cocom, E (2006), "Biomarkers and pollutants in the Nile Tilapia, Oreochromis niloticus, in four lakes from San Miguel, Chiapas, Mexico", *International Journal of Environment and Pollution*, v. 26, No 1-3, 129-141.]

[17] Hernández-Zamora, F, Roldán-Martín, A, Cruz-Colin, R, Castañeda-Briones, M, and Rodriguez-Vázquez, R (2006), "Evaluación de la toxicidad de un suelo contaminado con plaguicidas organoclorados antes y después del tratamiento de biolabranza", Memorias del XV Congreso Nacional de Ingeniería Sanitaria y Ciencias Ambientales, 24 al 26 de mayo del 2006, Guadalajara, URL: *http://www.femisca.org.mx /publicaciones/XVcongreso/XVCNIS141.pdf* (consulted on July 2008).

[18] Hyne, RV and Maher, W A (2003), "Invertebrate biomarkers: links to toxicosis that predict population decline", *Ecotoxicol Environ Saf*, v 54, No 3, 366-74.

[19] Jasso-Pineda, Y, Espinosa-Reyes, G, González-Mille, D, Razo-Soto, I, Carrizales, L, Torres-Dosal, A, Mejía-Saavedra, J, Monroy, M, Ize, A I, Yarto, M, and Díaz-Barriga, F (2007), "An integrated health risk assessment approach to the study of mining sites contaminated with arsenic and lead", *Integrated Environmental Assessment and Management*, v 3, No 3, 344–350.

[20] Jiménez, B, Rodríguez-Estrella, R, Merino, R, Gómez, G, Rivera, L, González, M J, Abad, E, and Rivera, J (2005), "Results and evaluation of the first study of organochlorine contaminants (PCDDs, PCDFs, PCBs and DDTs), heavy metals and metalloids in birds from Baja California, México", *Environmental Pollution*, v 133, 139–146.

[21] Kaputska, L A (2004), "Establishing Eco-SSL for PAHs: Lessons revealed from review of literature on exposure and effects to terrestrial receptors", *Human and Ecological Risk Assessment*, v 10, 185-205.

[22] LEXCORP Abogados (2008), "Emergency Official Mexican Standard NOM-EM-138-ECOL-2002, Establishing the maximum allowable limits for pollution of soils affected by hydrocarbons, site characterization and restoration procedures", URL: *http://www.lexcorp.com.mx/eng/Areas/PDF/Doc5eng.*pdf (consulted on June 2008).

[23] LGEEPA (Ley General del Equilibrio Ecológico y la Protección al Ambiente) (1998), *Ley General del equilibrio ecológico y la protección al ambiente (y disposiciones complementarias)*, XVI edición actualizada, Porrúa, México D.F., ISBN: 970-07-1346-6.

[24] Martínez-Tabche, L, Romero-Solís, M, López-López, E and Galar-Martínez, M (1999), "Efecto tóxico del DDT, Clordano y agua de la presa Ignacio Ramírez (México), sobre *Daphnia magna* (Crustacea: Daphnidae)", *Rev. Biol. Trop*, v 47, No.4, 681-690.

[25] Nagy, K A (2001), "Food requirements of wild animals: predictive equations for free-living mammals, reptiles, and birds", *Nutrition Abstracts and Reviews Series B: Livestock Feeds and Feeding*, v 71, No 10, 2R-31R.

[26] Pica-Granados, Y, Trujillo, G D, and Hernandez, H S (2000), "Bioassay standardization for water quality monitoring in Mexico", *Environ Toxicol*. v 15, No 4, 322-330.

[27] PROFEPA (Procuraduría Federal de Protección al Ambiente) (1998), *Memorias del grupo de trabajo sobre restauración de suelos contaminados,* Grupo de Trabajo sobre Restauración de Suelos Contaminados, Procuraduría Federal de Protección al Ambiente, México D.F.

[28] RAIS (Risk Assessment Information System) (2008), “Appendix F. food chain models for risk assessment”, Risk Assessment Information System, URL: *http://risk.lsd.ornl.gov/homepage/bjcor271/appf.shtml* (consulted on July 2008).

[29] Sample, B E and Arenal, C A (1999), “Allometric models for interspecies extrapolation of wildlife”, *Bulletin of Environmental Contamination and Toxicology*, No 62, 653-663

[30] Sample, B E, Opresko, D M and Suter II, G W (1996), *Toxicological benchmarks for wildlife: 1996 revision, ES/ER/TM-86/R3*, Risk Assessment Program, Health Sciences Research Division, Oak Ridge, Department of Energy, Office of Environmental Management, Oak Ridge.

[31] Santos-Medrano, G E, Ramírez-López, E M, Hernández-Flores, S, Azuara-Medina, P M, and Rico-Martínez, R (2007), “Determination of toxicity levels in the San Pedro River Watershed, Aguascalientes, Mexico”, *Journal of Environmental Science and Health. Part A, Toxic/Hazardous Substances & Environmental Engineering,* v 42, No 10, 1403-1410.

[32] SEMARNAT (Secretaría de Medio Ambiente y Recursos Naturales) (2002), *Guía para la presentación del estudio de riesgo ambiental Nivel 0. Ductos Terrestres*, Secretaría de Medio Ambiente y Recursos Naturales, México D.F., 50 p.

[33] SEMARNAT (Secretaría de Medio Ambiente y Recursos Naturales) (2006), *Guía técnica para orientar la elaboración de estudios de riesgo ambiental de sitios contaminados*, Secretaría de Medio Ambiente y Recursos Naturales, Dirección General de Gestión Integral de Materiales y Actividades Riesgosas, México D.F.

[34] Tannenbaum, L (2003), "Can ecological receptors really be at risk?, *Human and Ecological Risk Assessment*, v 9, No. 1, 5-13.

[35] TNRCC (Texas Natural Resources Conservation Commission) (2000), *Guidance for Conducting Ecological Risk Assessment at Remediation Sites in Texas. Draft Final*, Texas Natural Resources Conservation Commission, Texas, 267 p.

[36] TNRCC (Texas Natural Resources Conservation Commission) (2001), “Guidance for Conducting Ecological Risk Assessments at Remediation Sites in Texas”, RG-263 (revised), Toxicology and Risk Assessment Section, Texas Natural Resource Conservation Commission, Austin, URL: *http://www.tceq.state.tx.us/files/rg-263.pdf_4229898.pdf* (consulted on July 2008), 40.

[37] TPHCWG (Total Petroleum Hydrocarbon Criteria Working Group) (1997), *Volume 3, Selection of representative TPH fractions based on fate and transport considerations*, Amherst Scientific Publishers, Amherst.

[38] Travis, C C and Arms, A D (1988), “Bioconcentration of organics in beef, milk, and vegetation”, *Environmental Science and Technology*, v 22, No 3, 271-274.

[39] U.S.EPA (United States Environmental Protection Agency) (1998a), “Guidelines for Ecological Risk Assessment”, EPA/630/R-95/002F Final, Published on May 14, 1998, Federal Register 63(93):26846-26924, Risk Assessment Forum, U.S. Environmental Protection Agency, Washington D. C., URL: *http://www.epa.gov/ncea/raf/pdfs/ecotxtbx.pdf* (consulted on July 2008), 188 p.

[40] U.S.EPA (United States Environmental Protection Agency) (1998b), *Human health risk assessment protocol for hazardous waste combustion facilities. Volume Two. Peer review draft*, Office of Solid Waste and Emergency Response, U.S.EPA region 6, Multimedia Planning and Permitting Division Center for Combustion Science and Engineering, EPA530-D-98-001B, EUA:

[41] U.S.EPA (United States Environmental Protection Agency) (1999), "Screening level ecological risk assessment protocol for hazardous waste combustion facilities. Volume 2", Appendix A, EPA 530-D-99-001B, U.S.EPA Office of Solid Waste, URL: *http://www.epa.gov/epaoswer/hazwaste/combust/eco-risk/volume2/appx-a.pd* (consulted on July 2008).

[42] U.S.EPA (United States Environmental Protection Agency) (2005), "Guidance for Developing Ecological Soil Screening Levels, OSWER Directive 9285.7-55", Office of Emergency and Remediation Response, U.S. EPA, Washington, DC, URL: *http://www.epa.gov/ecotox/ecossl/pdf/ecossl_guidance_chapters.pdf* (consulted on July 2008), 1-6.

[43] Vázquez, A, Solís, E., Macedo, N, and Rosas, I (1986), "Influencia de la calidad del agua sobre la ocurrencia de *Daphnia pulex* en la Presa Jose Antonio Alzate y algunos aspectos de su pesquería", Contaminacion Ambiental, v 2, No 1, 39-56.

[44] Zapata-Pérez, O, Simá-Alvarez, R, Noreña-Barroso, E, Güemes, J, Gold-Bouchot, G, Ortega, A ,and Albores-Medina, A (2000), "Toxicity of sediments from bahía de chetumal, méxico, as assessed by hepatic EROD induction and histology in Nile tilapia *Oreochromis niloticus*" , *Marine Environmental Research,* v 50, 385-391.

In: Remediation of Soils and Aquifers
Editors: Luis G. Torres and Erick R. Bandala ISBN 978-1-60741-372-1

Chapter 7

INMOBILIZATION OF CONTAMINANTS IN SOIL AND REUSE AS IMPERMEABLE LAYERS FOR GROUNDWATER PROTECTION

Arturo Hernández-Antonio, Anne M. Hansen*, and Manfred van Afferden
Instituto Mexicano de Tecnologia del Agua

ABSTRACT

A technology for remediation of contaminated soils is presented. This method is based on the use of natural adsorbents that, due to their adsorption capacity, immobilize hydrophobic contaminants, reducing their leaching potential and thereby protecting groundwater. To test the technology, experiments were conducted using soil with cadmium added and mixing with natural adsorbents. The natural adsorbents were Calcic bentonite (5%), Activated carbon (5%), and Calcic bentonite:Activated carbon 9:1 (5, 7.5, 10, and 15 %). The most favorable moisture content, the permeability, and the immobilization of cadmium were determined for each experiment. The results confirm that Calcic bentonite immobilize cadmium more efficiently than Activated carbon. Cadmium was less mobile when the amount of natural adsorbents was increased, and 10% of the mixture of Calcic bentonite:Activated carbon 9:1 reduced the permeability by six orders of magnitude and adsorbed approximately 80% of cadmium, while addition of 15% of the mixture of natural adsorbents made the soil practically impermeable. The calculation for application of this technology in the immobilization of cadmium in contaminated soil is presented.

Keywords: Natural adsorbents, cadmium, contaminated soil, impermobilization, leaching reduction

---

* Corresponding author: ahansen@tlaloc.imta.mx.

INTRODUCTION

The main sources of pollutants in soil and groundwater can be classified as follows: a) Solid waste on the soil surface; b) Infiltration of contaminated water; c) Recharge of aquifers through injection of contaminated water through wells; d) Incorporation of natural contaminants through water-rock interactions in aquifers; e) Infiltration of agrochemicals; f)Transport of contaminants from buried waste like radioactive, chemical, latrine, and septic tanks; g) Intrusion of water from contaminated aquifers to non-polluted groundwater due to changes in direction and flow velocity, usually produced by overexploitation of the aquifers (Rabideau *et al.,* 2001). Common contaminants in soil are petroleum hydrocarbons spilled or deposited during operations such as extraction, refining, transport, and commercialization, contaminating soil with petroleum, combustoleum, gasoline, diesel, and turbosine (Eweis *et al.,* 1999).

The Mexican Ministry of the Environment and Natural Resources (SEMARNAT), has identified approximately 312 sites contaminated with hazardous waste in the country (SEMARNAT, 2007). Also, the Federal Office for Environmental Protection (PROFEPA) has elaborated criteria for restoration during environmental emergencies due to contamination of soil, being the following substances of main concern: 1) Crude oil, 2) Gasoline, 3) Combustoleum, 4) Diesel, 5) Sulfuric Acid, 6) Toluene, 7) Chrome, 8) Lead, 9) Arsenic, 10) Mercury, 11) Cadmium, 12) Nickel, 13) Barium, 14) Solvents, 15) Acrylamide, 16) Cyanide, and 17) Acrylonitrile (Calderón, 1999). In Mexico, the places where these wastes are most frequently deposited are bare terrains, industrial backyards, garbage dumps, canyons, sides of highways, municipal collectors, and water bodies. These deposits cause deterioration of air, surface water and groundwater qualities due to the migration of contaminants from soil to these final destinations. Another type of contamination is the deposition of garbage in open dumps where adequate treatment is often lacking thus generating leachates that may pollute surface water and aquifers.

Contaminated soil may be remediated through application of different physical, chemical, and biological methods. One solution is by not allowing contact with the environment and biota, interrupting the exposure pathways through encapsulation of the contaminated soil by construction of physical barriers, screens or sealing layers (Kleinken, 1997).

Another treatment technology consists in the immobilization of contaminants in the soil not allowing their environmental dispersion. This include vitrification of contaminated soil, mixing with chemicals such as hydroxides, sulfates, phosphates, paraffins, alcohols or fatty acids, to immobilize the contaminants through adsorption, precipitation, complexation, oxidation, reduction, polymerization or condensation.

Solidification by injection of cement, lime, gypsum or asphalts, represent other methods to immobilize contaminants in soils (Bölsing, 1995; Faschan *et al.*, 1993; Grathwohl and Einsele, 1994; Kruse, 1997; Kühnert *et al.,* 1993; TAUW Umwelt, 1996).

The permeability of soil may be reduced by addition of fine solids. The reduction in infiltration occurs due to the settling of the fine particles in the soil pores, reducing the mobility of water and diminishing the soil permeability. The concentration of mobile contaminants is also reduced due to adsorption on the surface of the fine solids. We will refer here to the combined impermeabilization and immobilization processes as "impermobilization".

This chapter describes the development of the technology of "impermobilization", based on the use of natural adsorbents such as Calcic bentonite and Activated carbon to immobilize contaminants in soil. Potential reuse of the stabilized soil as construction material for groundwater protection is explored.

MATERIALS AND METHODS

The experimental studies were carried out in the laboratories of Soil Mechanics and Hydrogeochemistry at the Mexican Institute for Water Technology. To determine optimal moisture contents that guarantee higher density or dry volumetric weight of the soil, compaction and permeability tests were carried out. Permeabilities of soil mixed with different amounts of Calcic bentonite, Activated carbon or mixtures of both natural adsorbents were also determined. Cadmium immobilization and transport were studied in soil columns under conditions of continuous flow.

Soil samples were combined with the natural adsorbents as presented in Table 1.

Table 1. Composition of the study solids.

Solid	Particle diameter (mm)	Amount (g)			
Activated carbon	0.075	12.50	18.75	25.00	37.50
Calcic bentonite	0.075	112.50	168.75	225.00	337.50
Gravel	9.5 to 19	1,187.50	1,156.25	1,125.00	1,062.50
Sand	0.3 to 0.425	1,187.50	1,156.25	1,125.00	1,062.50
Natural adsorbents (%)		5	7.5	10	15

a) Optimal moisture content

Optimal moisture contents of the solids were determined using of standard Proctor compactation tests (ASTM D698-2000), obtaining the higher dry volumetric weight. For this test, 2,500 g of solid mixtures were dried to constant weight in electrical furnace at temperature of 110 ± 5 °C during 24 h; MilliQ® water (Millipore Milli-Q Extra; Molsheim, France) was then added uniformly under continuously mixing to obtain moisture contents of approximately 8%. The prepared solids were located in the Proctor cylinder in three subsequent layers of 11-12 cm that were each compacted according to the procedure described in ASTM (2000a). Once packed, excess solid was eliminated with a spatula and the packed column was extracted. To determine the moisture contents, sub-samples were obtained from the upper, lower, and middle parts of the column. The procedure was repeated with moisture contents of 10, 12, and 15%. Maximum dry bulk density and optimal moisture content were determined by analyzing the resulting graphs of dry bulk density against moisture content.

b) Permeability

The hydraulic conductivities of the columns were determined with a flexible wall permeameter Solids with their respective optimal moisture contents were packed in cylinders (d=10 cm, h=14 cm) according to the method described in ASTM D5084-2000 (ASTM 2000b). Once extracted, resulting column diameters were reduced to 7 cm, covered with a latex membrane and placed in the permeameter (Ele International Tri Flex 2; Lake Bluff Illinois, USA) on a Whatman 40 filter (Kent, England) followed by a 0.5 cm high porous stone. Similar filters and porous stone were located in the upper part of the permeameter that was closed and located in a cell. Installed the column in the cell, the valves were closed and water was applied with a lateral pressure of 10 kPa until saturation. The amount of water permeated in the column was determined hourly and once stabilized, the permeability coefficient was determined.

c) Lixiviation Tests

A vertical and a horizontal column were tested for efficiency. In both columns, Whatman 40 filters (Kent, England) and porous stones were located in the top and the bottom of the columns as described above. In the vertical column (d=4.5 cm, h=9 cm) (Figure 1), the hydraulic load was applied by gravity. In the horizontal column (d=2.8 cm, h=3 cm) (Figure 2), the hydraulic load was applied by mechanical piston.

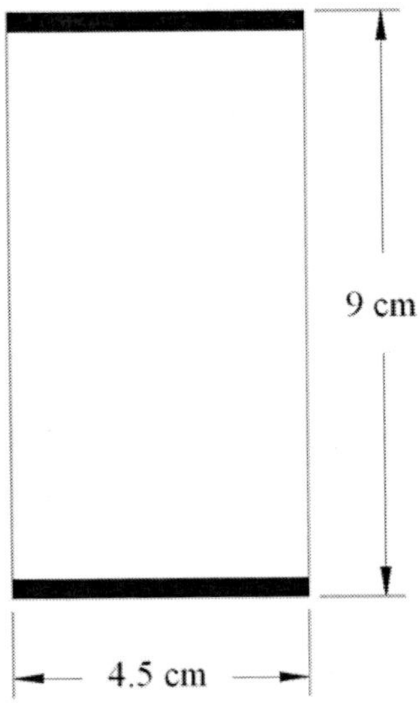

Figure 1. Vertical column.

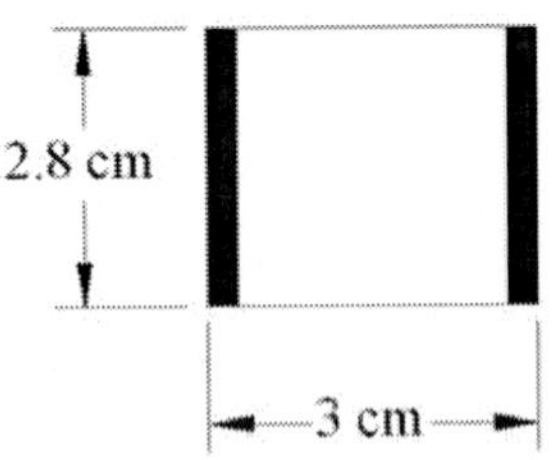

Figure 2. Horizontal column.

The characteristics of the columns are presented in Table 2.

Table 2. Characteristics of the columns.

Parameter	Column	
	Vertical	Horizontal
Input flow (m^3 s^{-1})	2.77×10^{-10}	1.07×10^{-10}
Porosity of the solid (%)	33	33
Compactation:		
Energy of compactation (kg cm cm^{-3})	2.9	2.9
Number of blows per layer	40	7
Number of layers	3	2
Height of fall of the piston (cm)	10	10
Weight of the piston (kg)	0.34	0.34
Volume (cm^3)	143.14	18.47
Permeability (m s^{-1})	1.6×10^{-7}	1.0×10^{-7}
Hydraulic gradient	13.2	8.4
Optimal moisture content (%)	12.5	12.5

d) Column Functioning

A flow of 0.016 mL min^{-1} or 0.5 mL h^{-1} was considered adequate for leaching tests. To examine the functioning of the columns and verify the existence of preferential flow and wall effect, two experiments were conducted (Figures 3 and 4, Table 3):

1. Applying of a pulse of sodium chloride in the upper part of the column, and
2. Distributing sodium chloride uniformly in the column.

Figure 3. Application of pulse of NaCl.

Figure 4. Uniform distribution of NaCl.

Table 3. Column functioning test conditions.

Parameter	Pulse of NaCl	Uniformly distributed NaCl
NaCl (mg L^{-1})	10,000	10,000
NaCl (mg)	27.5	63.6
Volume (mL)	2.75	6.36
Concentration of natural adsorbents (%)	10	10
Permeability (m s^{-1})	$1x10^{-7}$	$1x10^{-7}$
Number of layers	2	2
Number of blows per layer	14	14

Sodium chloride concentration in the leached solution was measured as electrical conductivity using a conductivimeter (Extech Instruments, Ohio, USA).

e) Cadmium Immobilization Experiments

Experiments to determine immobilization of cadmium were carried out by applying cadmium chloride with cadmium-109 (Isotope Products Laboratories; Burbank, California) as radioactive tracer uniformly to the solid with optimal moisture content, before packing the column (Figure 5). Leached solution was collected every 30 min and radioactivity was measured in a gamma counter (MINAXI 5530; Packard Instrument Co. Meriden, USA).

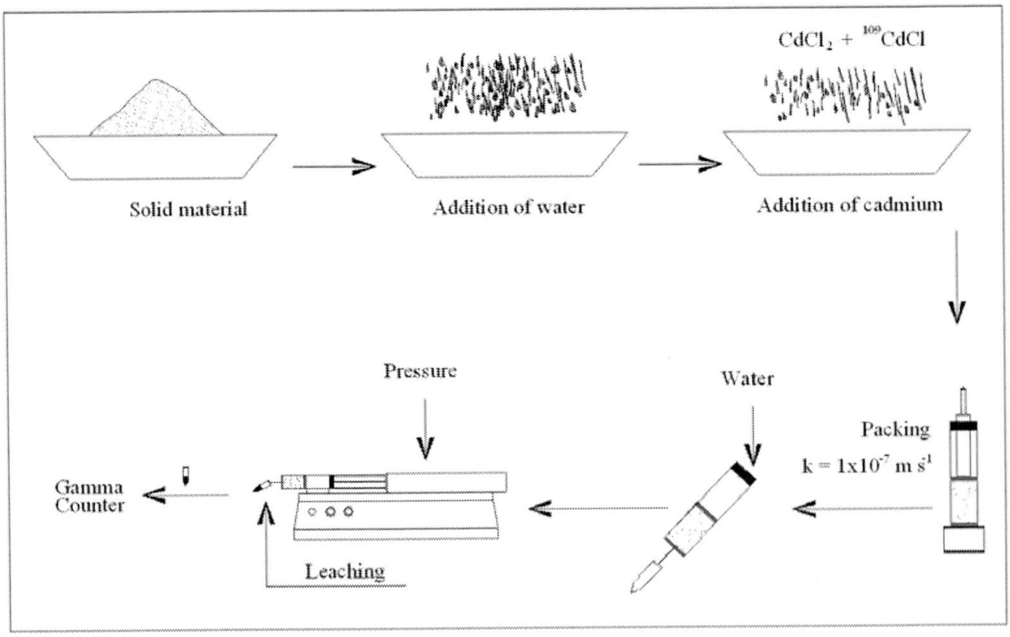

Figure 5. Schematic representation of the cadmium immobilization experiments.

Five experiments were conducted, one with the soil alone and four with different compositions and concentrations of natural adsorbents as described in Table 4. The leaching took place with a flow rate of 30 mL 30 h^{-1} and sampling was carried out with a frequency of 30 min.

Table 4. Experimental conditions during the cadmium immobilization experiments.

Parameter	Calcic bentonite:Activated carbon 9:1	Calcic bentonite or Activated carbon
Soil (g)	35	35
Natural adsorbent (%)	0, 5, and 10	5
Cadmium (mg kg^{-1})	851.21	851.21
Packing energy (kg cm cm^{-3})	2.9	2.9
Number of layers	2	2
Number of blows per layer	14	14

f) Application of the Technology

A scenario for application of the impermobilization technology is presented. It includes the remediation of two contaminated parcels whose characteristics are presented in Table 5, and potential reuse of the stabilized soil as construction material for groundwater protection.

Table 5. Characteristics of two contaminated parcels.

Parameter	Parcel I	Parcel II
Surface area (m^2)	300	600
Depth of the contaminated layer (m)	5.5	1.5
Volume (m^3)	1,650	900
Permeability (m s^{-1})	1.0×10^{-5}	1.0×10^{-5}
Depth of the groundwater table (m)	8	8

The concept of remediation of the area consists in extracting the contaminated layer in Parcel II and substituting this volume by clean soil. The extracted soil is then treated by the impermobilization technology and reused as a sealing layer in Parcel I to interrupt the infiltration of rainwater and leaching of contaminants to the aquifer. This strategy allows Parcel II, which is the largest of the two areas, to be cleaned up with no restrictions for its future use. Contamination in Parcel I will continue although infiltration of rainwater and leaching of contaminants to the aquifer will be interrupted by the sealing layer.

The sealing layer in Parcel I was considered as a standard system according to the guidelines established in Germany and the USA (TA-Siedlungsabfall, 1993; ASTM, 1995). These include a vegetation-soil layer, a drainage layer, a sealing layer, and an "equilibrium layer", that may be a biomembrane or clay minerals (Table 6).

Table 6. Stratification of the sealing layer in Parcel I.

Description	Depth (m)	k (m s^{-1})
Vegetation-soil layer	≥ 0.5	$1x10^{-5}$
Darinage layer	≥ 0.3	$1x10^{-3}$
Sealing layer (suelo saneado)	≥ 0.5	$1x10^{-9}$
Equilibrium layer	≤ 0.5	-
Contaminated soil	5.5	-

To demonstrate the efficiency of the impermobilization technology, the leaching of contaminants to the groundwater table is calculated before and after remediation. The mass of contaminants reaching the groundwater is:

$$L = Q_i \cdot C$$

Where

L = mass of contaminants (g s^{-1})
Qi = infiltration velocity (m^3 s-1)
C = concentration of contaminants (g m^{-3})

Q_i was calculated using the computer program HELP 3.0 (Hydrologic Evaluation of Landfill Performance) developed by the Environmental Protection Agency of the USA (Schroeder *et al.,* 1994). This program allows modeling hydrological processes such as runoff, infiltration, evapotranspiration, drainage, and percolation or leaching. HELP was developed for planning landfills and we applied it here to evaluate the efficiency of sealing layers and to estimate the leaching of water and contaminants to the groundwater table.

According to the runoff curve "SCS Runoff Curve Number" (Rawls *et al.,* 1993), for the modeling of runoff, the natural condition (NC) of the area was defined as NC = 88.1. Daily climatological data (precipitation, radiation, and temperature) were simulated based on information obtained from the climatological station 00023019of Puerto Morelos, Quintana Roo, Mexico, from 1971 to 2000 (SMN, 2008).

Results Y Discussion

Results of optimal moisture contents of soil mixed with Calcic bentonite:Activated carbon 9:1 vary between 9 and 14 % (Figures 7 and 8). Increasing the amount of natural adsorbents results in higher demand of water and in lower dry volumetric weight. These results were applied to establish the optimal moisture content, guaranteeing maximum dry volumetric weight of the solid mixtures to be used in the cadmium immobilization experiments.

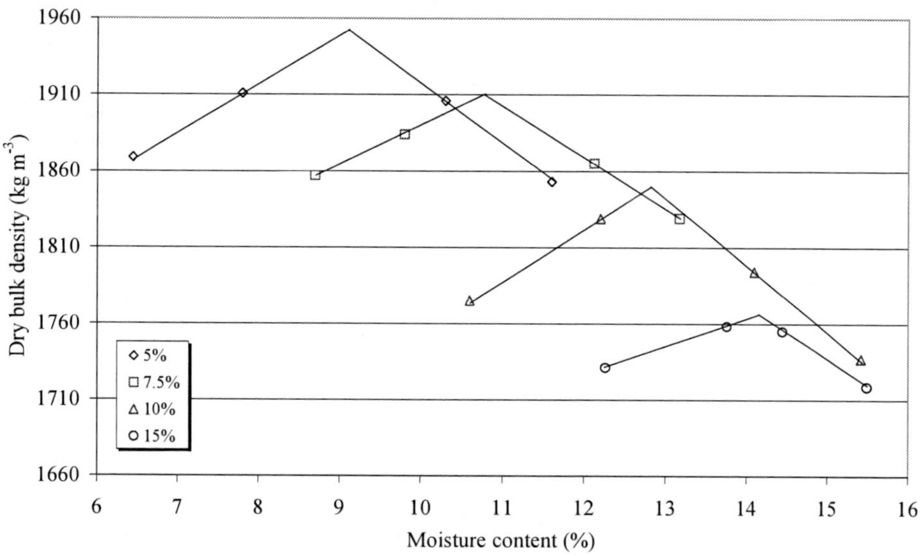

Figure 7. Determination of optimal moisture content in soil with different amounts of Calcic bentonite:Activated carbon 9:1.

Results of permeability of treated soil show the decrease in this parameter when the amount of natural adsorbents mixed with the soil sample is increased (Table 6). Permeabilities of less than $1x10^{-8}$ m s^{-1} are observed when the soil is mixed with 10% of natural adsorbents or more.

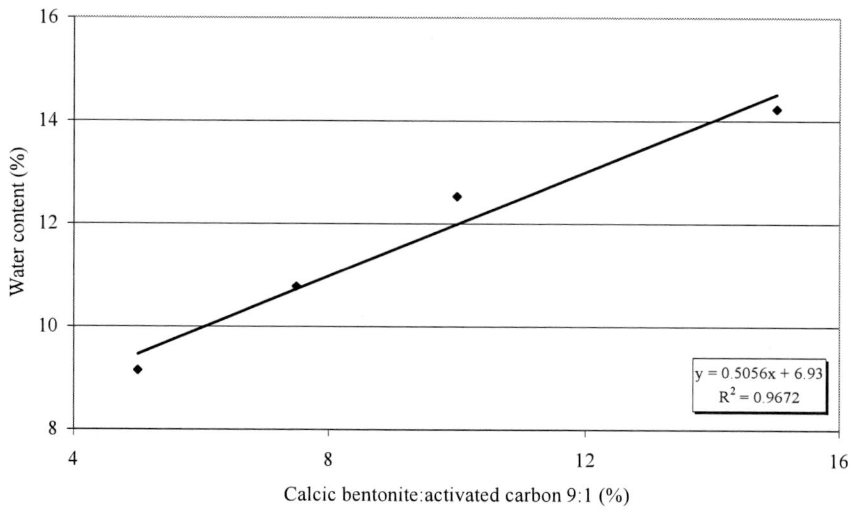

Figure 8. Optimal moisture content of the mixtures of soil with natural adsorbents.

Table 6. Determination of permeability.

Natural Adsorbent (%)	Packing Energy (kg cm cm^{-3})	K (m s^{-1})
5.0	2.90	2.96x10^{-7}
7.5	3.60	6.12x10^{-8}
10.0	4.79	8.98x10^{-9}

Results of the functioning of packed columns are presented in Figure 9. Flow rate in the vertical column was high during the first hours of the experiment after which the flow reduced to approximately 0.01 mL min^{-1}, value below the established experimental minimum of 0.5 mL 30 min^{-1}. Flow rate in the horizontal column was more constant and stabilized at around 0.027 mL min^{-1} or 0.8 mL 30 min^{-1}, that fulfilled the experimental minimum requirement. Therefore the horizontal column was used in the subsequent experiments.

In experiments where sodium chloride was applied as a pulse on the top of the column, leaching occurred as a peak that reached a maximum concentration after 4 h (Figure 10). The total amount of sodium chloride leached from the column during 43 h of experimentation, represented 94% of the total amount added to the column. The presence of a single maximum in concentration of leached sodium chloride confirms a good operation of the column, without preferential flow or wall effect due to cavities between the solid and the wall of container. This column is therefore adequate for contaminants transport experiments.

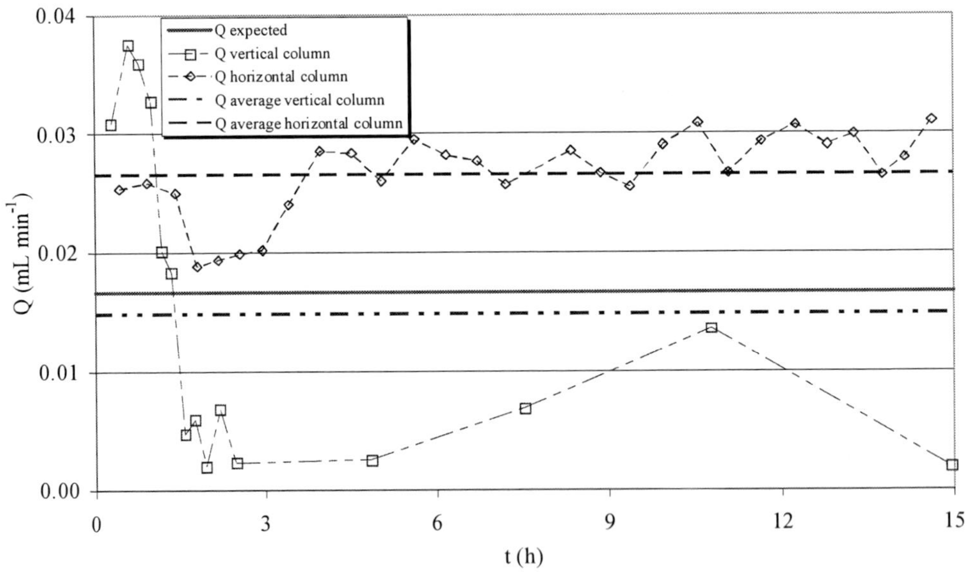

Figure 9. Flow in soil columns with 10% Calcic bentonite:Activated carbon 9:1.

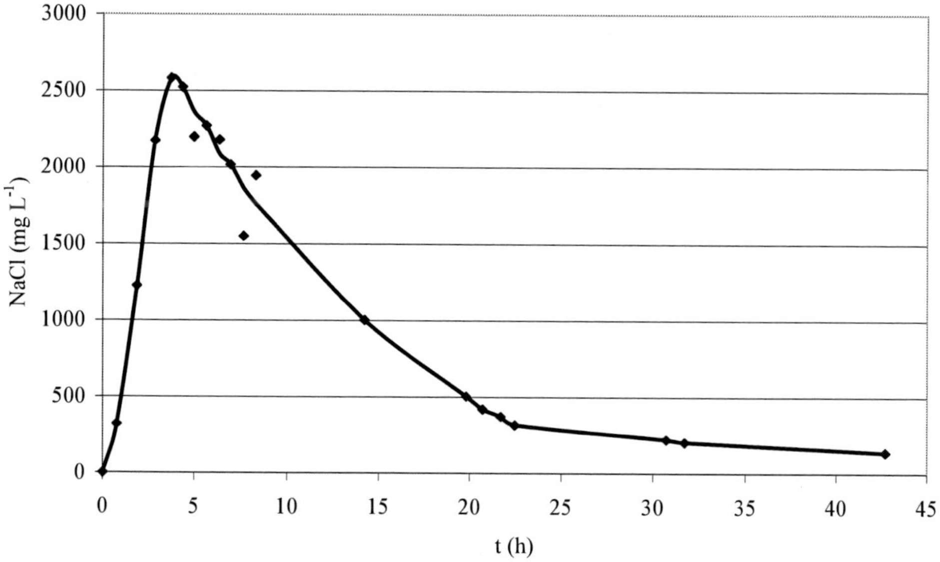

Figure 10. Transport of a pulse of NaCl 10,000 mg L^{-1}, 10% natural adsorbents, Q=0.5 mL 30 min^{-1} in horizontal column.

In experiments with sodium chloride uniformly distributed in the column, a single maximum of leached concentration was observed at the beginning of the experiment (Figure 11). The total amount of sodium chloride leached from the column during 12 h of experimentation, represented 90% of the total amount added to the column. This system was therefore also suitable for contaminants leaching experiments.

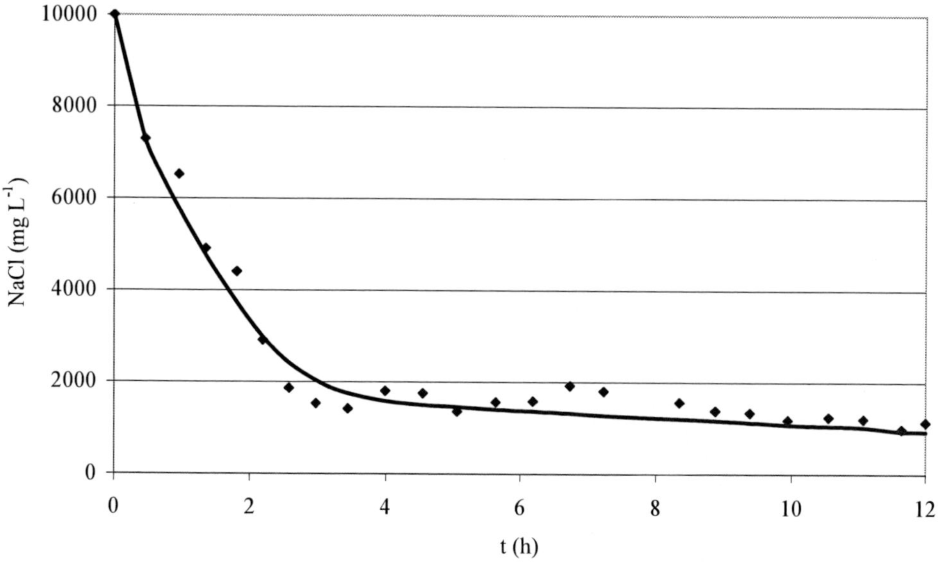

Figure 11. Transport of uniformly distributed NaCl 10 000 mg L^{-1}, 10% natural adsorbents, Q=0.5 mL 30 min^{-1} in horizontal column.

The results of transport of sodium chloride indicate that column experiments can be carried out in both systems (as a single pulse or with uniform distribution in the solid of the column). The system with uniform distribution was selected for experiments with cadmium, because it represents better the situation of contaminated soil whereas the pulse system would represent a situation where contaminated water infiltrates in the soil.

Cadmium Immobilization Studies

Addition of 10% of natural adsorbents with the soil caused an immobilization of approximately 80% of cadmium in 10 h (Figure 12 and Table 7) whereas 5% of adsorbents immobilized only 53% of the metal in 30 h. With no addition of natural adsorbents, 75% of cadmium is leached in 5 min. These results clearly demonstrate the benefit of addition of the natural adsorbents in reducing the risk of leaching of cadmium in contaminated soil.

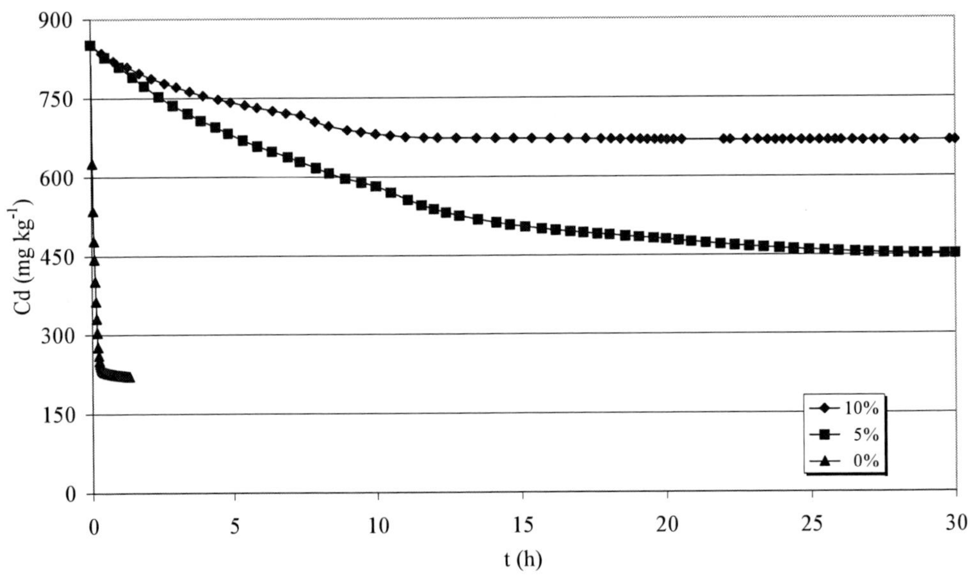

Figure 12. Immobilization of cadmium in soil with different concentrations of Calcic bentonite:Activated carbon 9:1.

Table 7. Immobilization of cadmium in soil with different concentrations of Calcic bentonite:Activated carbon 9:1.

Natural adsorbent (%)	t (h)	Immobilized cadmium (mg kg^{-1})	Immobilized cadmium (%)
0	0.08	212	25
5	30	451	53
10	30	681	80

Addition of 5% Activated carbon (Table 8 and Figure 13), caused an immobilization of 41% cadmium in the column after 30 h of experimentation while 5% Calcic bentonite resulted in 65% immobilization of the metal in the same period of time, showing the higher capacity of Calcic bentonite to immobilize cadmium than Activated carbon.

Table 8. Immobilization of cadmium in soil with 5% natural adsorbents (t = 30 h).

Natural adsorbent	Immobilized cadmium (mg kg^{-1})	Immobilized cadmium (%)
Activated carbon	348.99	41
Calcic bentonite	553.29	65

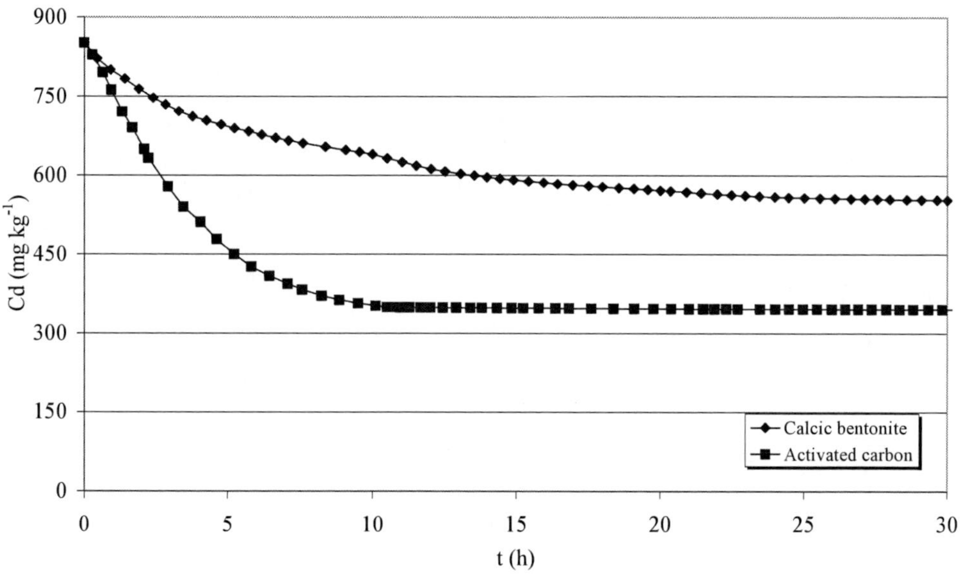

Figure 13. Immobilization of cadmium in soil with 5% natural adsorbents.

Infiltration before Remediation

Yearly water balance for the contaminated parcels before treatment is presented in Table 9 and Figure 14. Average surface runoffs in Parcels I and II were estimated in is 8.5% of precipitation, and evapotranspiration was estimated in 68% of precipitation. Infiltration represents 22% of the rainfall in Parcel I and 23% in Parcel II.

The proposed remediation scheme results in the total clean-up of parcel II as contaminated soil has been completely removed with no risk of future groundwater contamination. Soil in Parcel I, on the other hand, is still contaminated but the infiltration of rainwater is drastically reduced by the construction of a sealing layer mixing the excavated material from Parcel II with 10% Calcic Bentonite:Activated Carbon 9:1.

Table 9. Water balance in Parcels I and II before remediation.

Parameter	Total per year	
	Volume in Parcel I ($m^3 300 m^{-2}$)	Volume in Parcel II ($m^3 600 m^{-2}$)
Precipitation	340	681
Runoff	29	58
Evapotranspiration	231	462
Drainage	4	1
Infiltration	76	160

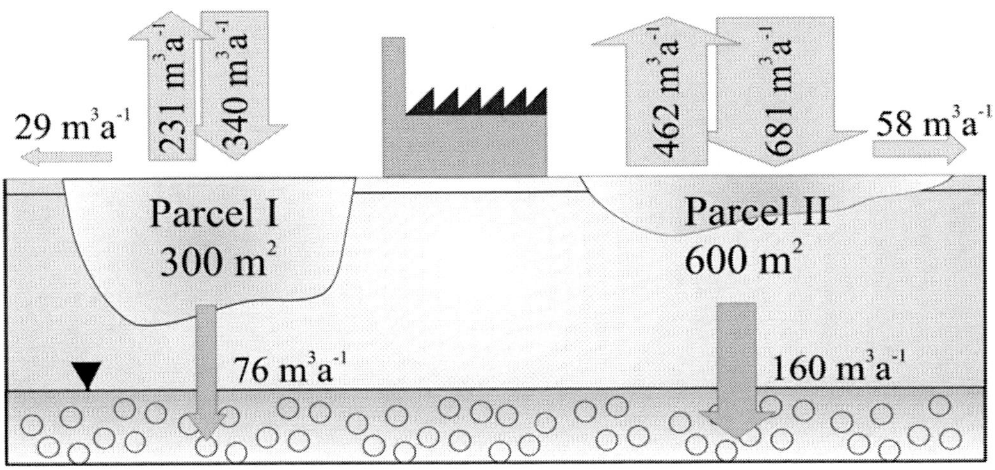

Figure 14. Water balance in the parcels before remediation.

INFILTRATION AFTER TREATMENT

Modeling of the water balance in Parcel I with the sealing layer (Table 10 y Figure 15) resulted in a reduction in infiltration from 22 to 1.5% of the rainfall. This means that the amount of cadmium leached to groundwater is also reduced to less than 7% of the amount prior to the remediation.

Table 10. Water balance in Parcel I after remediation.

Parameters	Total per year
	Volume (m^3 300 m^{-2})
Precipitation	340
Runoff	100
Evapotranspiration	231
Drainage	4
Infiltration	5

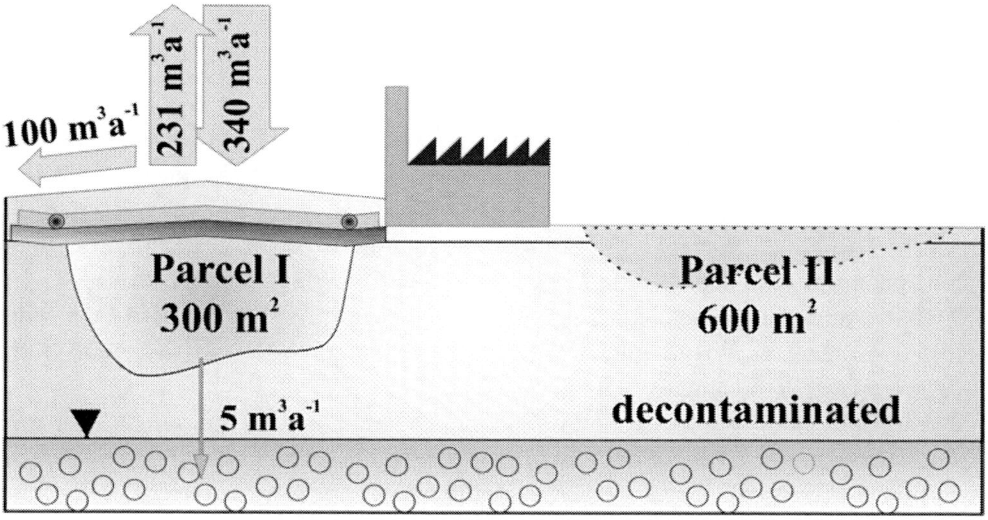

Figure 15. Water balance in the parcels after remediation.

CONCLUSIONS

This study demonstrates that the technology of impermobilization using natural adsorbents as Calcic bentonite and Activated carbon present an attractive option for construction of impermeable layers in the protection of groundwater. The results presented here confirm that the addition of 10% of a mixture of Calcic bentonite:Activated carbon 9:1 to sandy soil may reduce the permeability by up to six orders of magnitude. The adsorption capacity of the mixture of natural adsorbents used in this study, reduced cadmium concentrations with up to 80% by adsorption, being Calcic bentonite the main adsorbing agent for this metal. Numerical modeling with HELP (Schroeder *et al.*, 1994) demonstrated the potential of reuse of the treated soil as sealing layers, and its potential for reducing in 93% the volume leached to groundwater. The impermobilization technology can be applied in treatment of contaminated soil. The resulting material may be reused like material of compaction in the construction of sealing layers, systems of isolation like system of encapsulation, as well as in left sites and of this form to diminish the risk of infiltration of movable polluting agents towards underground waters.

ACKNOWLEDGMENTS

M.C. Carlos Gutiérrez Ojeda, for the support during the project development.

M.I. Isaac Bonola Alonso, for his advice during the experiments and review of the manuscript.

REFERENCES

[1] ASTM (American Society for Testing and Materials). 1995. Standard Guide fore Design of a Liner System for Containment of Wastes. In: ASTM (ed) Annual Book of ASTM Standards. Environmental Assessment; Hazardous Substances and Oil Spill Responses; *Waste Management,* Vol 11.04.

[2] ASTM (American Society for Testing and Materials). 2000a. Standard Test Methods for Laboratory Compaction Characteristics of Soil Using Standard Effort, D698.

[3] ASTM (American Society for Testing and Materials). 2000b. Standard Test Methods for Measurement of Hydraulic Conductivity of Saturated Porous Materials Using a Flexible Wall Permeameter, D5084.

[4] Bolsing, F. 1995. Method of immobilizing contaminants in the soil or in materials similar to the soil. United States 5413616. (Cl.71/27; C05G3/04), 05 May 1995. Appl. 103878, 9 Aug 1993. 5 p.

[5] Calderón J. L. 1999, PROFEPA reconoce esfuerzo de académicos. Teorema, Vol 22, pp 56-57., México, D.F., México.

[6] Eweis, J.B., Ergas, S.J., Chang, D.P.Y., and Schroeder, E.D. 1998. Bioremediation Principles. New York: McGraw-Hill. 296 p.

[7] Faschan, A., Tittlebaum, M. and Cartledge, F. 1993. Nonionic organic partitioning onto organoclays. Hazardous waste and hazardous materials, 10 (3). Liebert, New York. pp. 313-322.

[8] Grathwohl, P. and Einsele, G. 1994. Vergütung mineralischer Dichtungen zum Rückhalt organischer Schadstoffe. Offenlegungsschrift (Anmeldetag: 15.01.93) DE 4300915 A1. Deutsches Patentamt München.

[9] Kleinken A. 1997.Auswertung der Erfahrungen aus durchgeführten Sicherungsverfahren bei Altlasten. In LUA NRW (ed) Materialien zur Altlastensanierung und zum Bodenschutz. Band 3. Essen, Germany.

[10] Kruse, K. 1997. Process for immobilizing organic and inorganic pollutants in a contaminated soil material on a remediation site. United States 5651831. (Cl.134/18; B08B7/04), 17 Mar 1994. Appl. 232112, 3 Sep 1993. 5 p.

[11] Kühnert M., Fuchs V., Golbs S., Knauf H. 1993. Verfahren zur Herstellung von Huminsäure-Metall-Verbindungen, deren Verwendung als Adsorptions- und/oder Entgiftungsmittel bzw. als Spurenelementquelle. DE 4129872A1. (IPC.A61K35/10; C07G17/00), 11 Mars 1993. Offenlegungsschrift 4129872, 09 Sep 1991. 5 p.

[12] Rabideau, A. J., Van Benschoten, J. E., Khandelwal, A., Patel, A., and Repp, C. R. 2001. Sorbing vertical barriers. In Physicochemical Groundwater Remediation, Smith, J. A. and Burns, S. E. (eds.), Kluwer Academic/Plenum Publishers, 115-138.

[13] Rawls, W.J., L.R. Ahuja, D.L. Brankesiek, and A. Shirmohammadi, 1993. Infiltration and soil water movement. In D.R Maidment (ed). Handbook of Hydrology, Mc Graw-Hill, 5.1-5.51., NY.

[14] Schroeder, P., Dozier, T., Zappi, P., McEnroe, B., Sjostrom, J. and L. Peyton (1994). "The Hydrologic Evaluation of Landfill Performance (HELP) Model: Engineering Documentation for Version 3." EPA/600/R-94/168b, US. Environmental Protection Agency, Risk Reduction Engineering Laboratory, Cincinnati, OH. 116 p.

[15] SEMARNAT (Secretaría de Medio Ambiente y Recursos Naturales). 2007. Primer informe de labores. www.semarnat.gob.mx/queessemarnat/Pages/primerinformedelabores.aspx. Consulted on August 18, 2008. 159 p.

[16] SMN (Servicio Meteorológico Nacional) 2008. Normales Climatológicas 1971-2000, Station 00023019 Puerto Morelos, Quintana Roo, Mexico. Comisión Nacional del Agua, http://smn.cna.gob.mx/productos/normales/estacion/qroo/NORMAL23019.TXT, Consulted on November, 2008. 159 p.

[17] TA Siedlungsabfall, Technische Anleitung zur Verwertung, Behandlung und sonstigen Entsorgung von Siedlungsabfällen (1993). *In* Bundesanzeiger, Verlags-Gesellschaft mbH, Köln.

[18] TAUW Umwelt GmbH. 1996. Immobilisierung von Schadstoffen in Altlasten. In Sächssisches Landesamt für Umwelt und Geologie (ed) Materialien zur Altlastenbehandlung 1/1996.

In: Remediation of Soils and Aquifers
Editors: Luis G. Torres and Erick R. Bandala
ISBN 978-1-60741-372-1

Chapter 8

HYDROGEOLOGICAL CRITERIA FOR MONITORING PROGRAMS DESIGN IN CONTAMINATED AQUIFERS

J.A. Hernández-Espriu, J.A. Reyna-Gutiérrez, Luis E. Marín
Facultad de Ingeniería e Instituto de Geofisica-UNAM

ABSTRACT

Monitoring of contaminated aquifers must be carried out based first on sound hydrogeological criteria, and secondly on budget. Designing effective monitoring systems must address the following issues: (1) development of the conceptual model, (2) spatial distribution of the monitoring network, frequency and type of sampling, (3) design of wells and/or piezometers, (4) aquifer instrumentation and its application in long term programs, (5) and optimization of programs to assure quality through time while reducing costs. The aspects of each point are analyzed from the theoretical foundation and data from four Case-Studies area taken to illustrate the relevance of each criteria.When the hydrogeological features of the site, the biogeochemical conditions of the aquifer and the properties of the contaminant are taken into account, monitoring programs are the most important tool for (a) complementing the environmental assessment, (b) improve the conceptual model, and (c) generate indicators to determine the success of an environmental prevention, contention or remediation project.

Keywords: Geology, hydrogeology, polluted aquifers, monitoring program, contaminant plume.

1. INTRODUCTION

The international experience in environmental research and studies captured in specialized literature in regards to aquifer monitoring provides diverse guides to carry out programs in a more efficient way, but few are the documents that explain the technical criteria required to design a successful monitoring program that allows to optimize the state of the art in understanding the conceptual model of the contaminated site based on the Mexican experience and environmental problematic.

The actual situation surrounding contaminated sites management in Mexico reflects a problematic from the conceptual point of view, where only (in general) one part of the problem is addressed. Most of the times only soil contamination is addressed and due to a lack of understanding, groundwater, the most dynamic element of contaminant transport, is overseen.

This vision has made contaminated sites management to be addressed partially, sometimes even using misleading fundamental concepts such as thinking that an Environmental Assessment Phase II is achieved successfully by taking some soil samples (and its later analysis in a lab) or that site remediation consists in removing a few tons of contaminated soil for its later confinement or incineration.

This document is intended to bring into the picture that contaminated aquifer monitoring before, during and after the remediation project is much more complicated than taking a pair of groundwater samples. In this sense, a monitoring program on mid and long term is a fundamental tool that when well designed and founded on firm hydrogeological bases allows better understanding the conceptual model, complementing the characterization and determining the success of a contention, prevention or remediation system surrounding contaminated sites management.

2. Essential Elements of a Groundwater Monitoring Program

There are near 200,000 contaminated sites in the USA that are being monitored. Each site has an average of 10 monitoring wells and each well is sampled an average of three times per year. Each monitoring event (sample, analysis and report) costs around $ 2,500 per well (Nyer, 1988). According to this number, each site costs $ 75,000 per year, adding up to a grand total of $ 15,000 million yearly on contaminated sites monitoring.

In Mexico, although there are not published records of the monitoring costs, authors experience show that the total expense of monitoring exceeds 50% of the total cost of the project. Because of the high expenses derived from monitoring, it is of vital importance to review the main aspects in designing a monitoring program to assure the temporal quality of the collected data while reducing costs.

The recommendation on which elements must be considered thoroughly for implementing a contaminated aquifer monitoring program are: (1) understanding of the hydrogeological conceptual model as a starting point prior to any further analysis, (2) selection of parameters and analytes as contaminant indicators, (3) installation of piezometers and monitoring wells, from the selection of the drilling process, constructive design (controlled by the geological environment), development and purging strategies, (4) instrumentation of aquifers and its application on long term programs, (5) optimization of long term programs, and (6) budget considerations.

These criteria will be analyzed and technical guides will be given on how to design monitoring programs according to the elements mentioned above.

3. Hydrogeological Criteria for Monitoring Programs Design

3.1. Conceptual Frame

3.1.1. Geology

The starting point to design a monitoring network for contaminated aquifers is, without doubt, understanding the conceptual model of the site. The minimal considerations that must be known to understand the conceptual model can be grouped mainly into geological and hydrogeological aspects. Hydrogeology controls subsoil water movement, and geology is present in all factors that allow to depict the geological environment (soils, sediments, and rocks) in which the groundwater flows and its stored, such as lithology, structural features, mineralogy and spatio-temporal distribution of lithofacies (Sara, 2003).

Geological data required to establish the first preliminary sites for the installation of monitoring wells up-gradient and down-gradient, shall be oriented so that field investigations and subsequent analysis provide the following information:

Superficial Geology and Litho-stratigraphy

1. The type of geological environment through which the water flows such as porous, fractured, karstic, mixed, or assimilated (term used for fractured or karstic media that behaves as porous).
2. Spatio-temporal distribution of the outcropping rocks placed on a regional and local map.- (geological cartography). The regional geological map can be produced from a combination of previous maps, photogeological interpretations and field point verifications in order to analyze the dominant geological events and their influence on contaminated materials in the study zone. Geological cartography must be fitted on a base detailed topographic map, specifically made for the site. Finally, the outcome cartography must show all the geological, environmental and external attributes that could be affected by contaminants in the zone.
3. Discontinuities and structural geology characterization (faults and fractures) that control the subsurface flow and contaminant dispersion.- Especially in fractured terrenes, it will be essential to establish the control of tectonic structures (orientation, inclination, distribution, density, fill, openness, and kinematics) that can alter groundwater flow mainly because: (A) discontinuities may represent impermeable barriers that modify flow paths when they are filled by secondary carbonates, oxides or clay minerals, for example, in shear zones, and (B) discontinuities may represent focal zones of secondary porosity with highly hydraulic conductivity that generate local preferential flow paths. In some case certain structural patterns may connect hydraulically two aquifers where contaminated water interacts with good quality water or well vertical ascending flows may be induced so that contaminated water discharge in superficial water bodies may occur. Thus, it is imperative to take into account and consider this structural control as fundamental criteria during the design of monitoring networks.

Subsurface Geology

1. Distribution of geologic units in the subsoil. Multiple geological sections must be developed in different directions, to identify vertical and horizontal changes of facies acting as low permeability zones or as perched aquifers vulnerable to contamination. Additionally, rock thickness and saturated thickness of contaminated units must be identified.

3.1.2. Indirect Survey Methods

Indirect methods are a complementary tool to interpret the distribution of geological units in the subsoil. Even though for environmental studies, indirect methods are focused to gas and organic vapors monitoring in the vadose zone, geophysical prospecting in geo-environmental applications (much undeveloped in Mexico) can help to complete the sub-superficial portrayal.

There are many geophysical methods, each one with its unique principle, theoretical foundation, methodology and application according to the objectives of each project. In table 1 are summarized the most common geophysical methods including the physical property of measurement and its application field.

From table 1, it can be observed that most of the methods may contribute to site assessment, but electric and electromagnetic methods, because resistivity response of water can be very well tracked, are the most suitable ones to solve most of the problems related to hydrogeology with a better cost-benefit relationship (Casas, 2000). Geophysical methods help determine favorable zones of groundwater accumulation whether they are fresh, brackish or contaminated. For each particular case, an analysis must be made to determine which one or which ones are the most convenient to be carried out as some times the combined use of techniques allows to accomplish better results. In this regard, the depth of investigation, type of array and technique selection is of special interest.

Next is presented data from geophysical prospecting (Case-Study 1) in order to show some of the practical applications that electromagnetic methods have in groundwater assessment. The prospection was carried out in the Yucatan Peninsula, 70 km far northeast from Chetumal in a region known as "Costa Maya". The objective of this study was to determine the degree of contamination of the fresh water aquifer due to marine water salt intrusion and to establish with help from indirect methods the location of the fresh water – salt water interface (mixing zone) for a later drilling and water supply system program in the area. Based on the needed depth of investigation, the lithology and hydrogeological conditions, it was determined that Transient Electromagnetics (TEM), also known as Time Domain method, was to be applied. TEM consists of circulating a current and stopping it abruptly (pulses) through a transmitting coil (the depth of investigation is directly proportional to the area enclosed by the coil) in order to induce currents to the subsoil according to Maxwell's equations. These induced transient currents generate a secondary field which propagates with a velocity proportional to the resistivity of the subsoil. A receiver at the surface measures these temporal variations of the secondary field.

Results of a TEM sounding are presented as curves of apparent resistivity variations as a function of time, and its corresponding interpretation is done in an analogue way to that of the Vertical Electrical Soundings (VES) in which variations of the apparent resistivity are associated to geoelectrical units.

Table 1. Principal geophysical methods and general applications.

Geophysical exploration method	Physical property	Application field
		Cavity and regional geological structures detection
Gravimetric	Mass (density)	Geometric watershed analysis
		Basement assessment
Electrical		Groundwater prospecting
		Cavities and Karstic Systems assessment
Vertical electrical sounding (VES)	Resistivity	Clay lenses and confining layers thickness determination
Electrical tomography		Aquifer saturated thickness determination
Induced polarization	Chargeability	Fresh – salt water interface detection (coastal aquifers)
Electromagnetic	Resistivity	Deep groundwater prospecting
Transient Electromagnetics (TEM)	Magnetic susceptibility	Electrical anisotropy determination
Magnetotelluric	Hydrogen content	Alluvial covers thickness determination
Magnetic Resonance (MRI)	Dielectric constants	Water content determination
Ground Penetrating Radar		Environmental assessment, archeology, ice thickness and pipeline detection
Magnetometric	Magnetization	Mineral exploration
Seismic		
Refraction	Seismic waves transit time	Regional geological structures assessment
Reflection		Regional oil field exploration

Modified from (VanBlaricom, 1992*)*.

The work for this study consisted in various field campaigns in which 20 TEM sounds were carried out distributed across the length and breadth of and area formed of an alternation of calcarenites and heterogeneous fossiliferous Wackestone, Packestone and Grainstone limestone belonging to the Carrillo Formation from the Miocene-Pliocene (the aquifer of the study area). Sirotem-S (Mark 3) equipment with an array known as "Coinciding Loop" and a 50 x 50 m coil area was used to reach an investigation depth of 200 m.

Every material, depending on its properties (hardness, fracturing, fluid saturation and salinity, among others), has a characteristic resistivity response that is well enclosed within a range and documented on literature, see figure 1. Since brackish-salt water has a low resistivity response compared to that of fresh water or rocks, this response along with particularities analyzed from the geological, hydrogeological and hydrogeochemical conceptual model were taken to make an integrated interpretation of the data in order to determine the fresh water – salt water interaction, establish the location of the saline interface and determine the degree of contamination due to brackish-salt water.

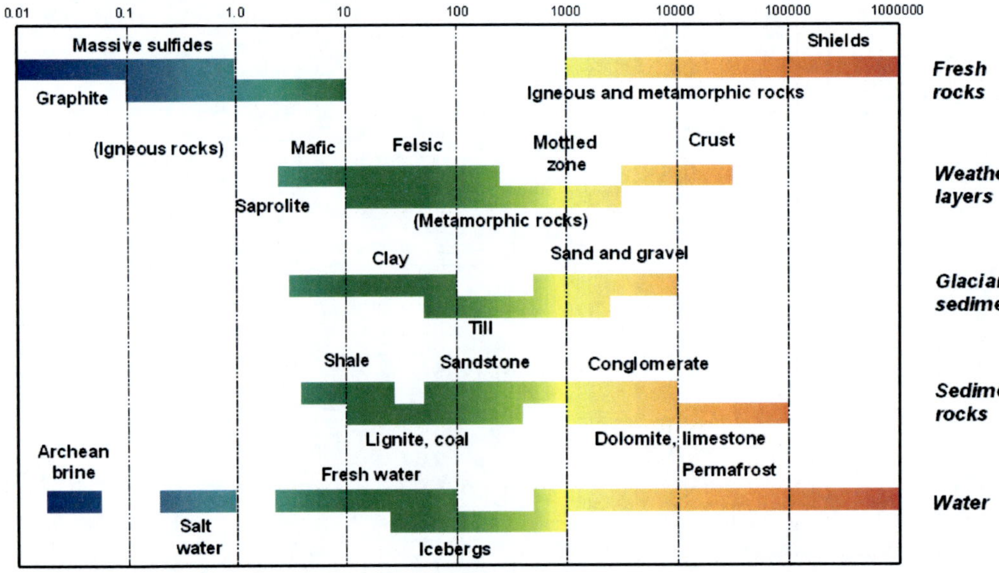

Figure 1. Resistivity response of different geological materials. Modified form (Arango, 2005).

For illustrative purposes, section L1 is presented in figure 2. This section is a 12.5 km transect with 13 TEM sounds oriented in a N25°E line (parallel to the coastline).The interpretation of L1 together with other 3 sections (which are not shown) allowed to depict the subsoil into 4 geoelectrical units, called UE1, UE2, UE3, and UE4 (the main features of each unit are described in table 2).

As it can be seen from figure 2, the first 20 m of the subsoil consists of a calcareous unit with resistivities from 26 to 104 Ω·m, indicative of an aquifer containing a mixture of fresh water with slightly brackish water. Thus a low degree of contamination. From this depth on, resistivity values decrease notably ranging from 2 to 11 Ω·m, indicative of unit containing a conductive fluid which means a mixture of brackish with slightly salty water, at this depth, 20 - 40 m, a significant degree of contamination can be established and without doubt a progressive decrease of water quality with depth increase.

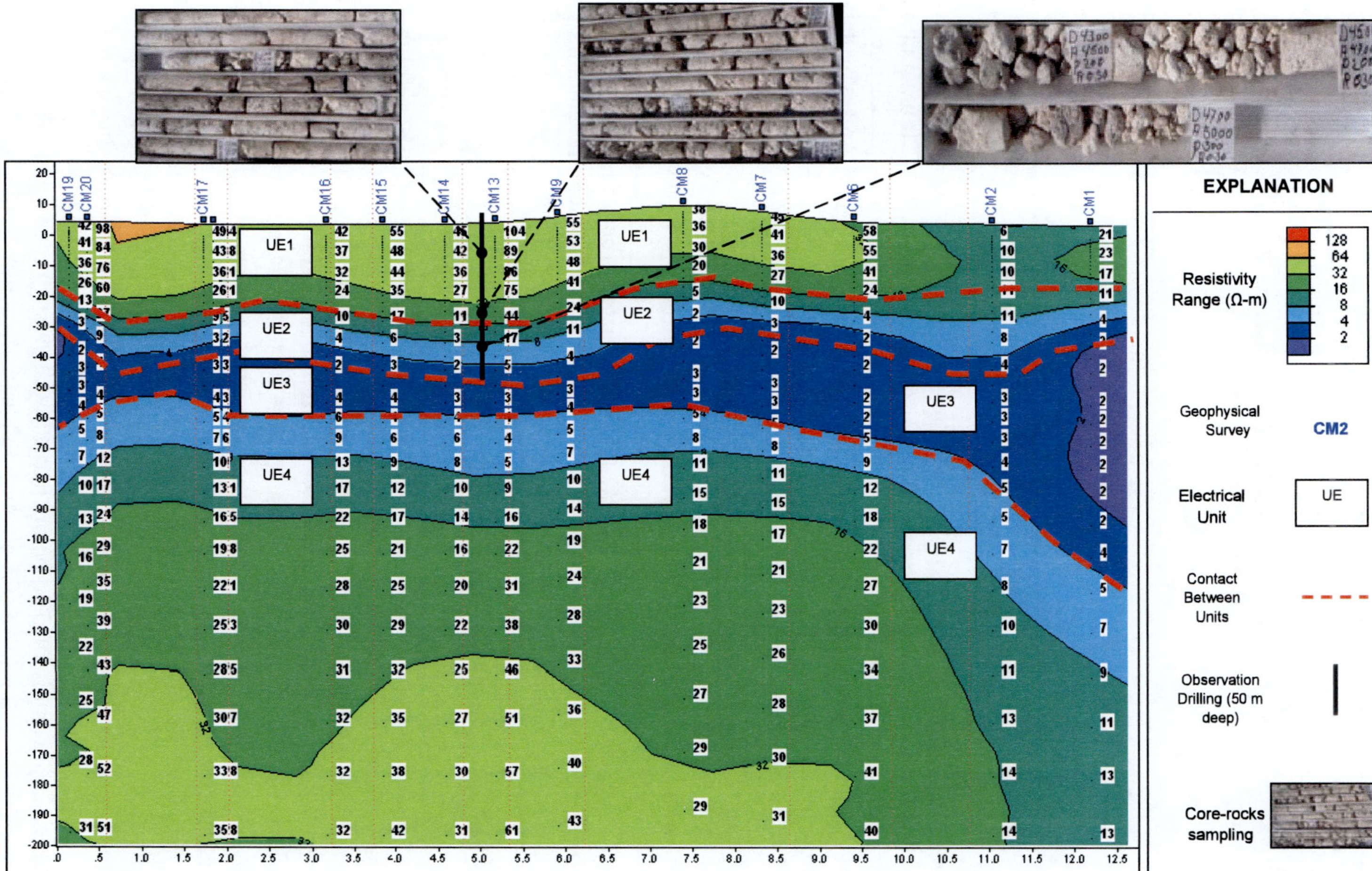

Figure 2. L1 resistivity section, the horizontal axis shows the distance in [km] and the vertical axis the depth in [m].

Table 2. Electrical interpreted units description.

Electrical Unit	Resistivity interval (Ω-m)	Geological correlation	Likely thickness (m)	Hydrogeological correlation and expected water quality
UE1	26 - 104	Calcarenite	18 – 25	Vadose zone within the first 2 m, fresh – slightly brackish water aquifer
UE2	2 – 11	Highly disolved fosiliferous limestone	12 – 20	Brackish - slightly salty water aquifer
UE3	2 - 5	Very highly dissolved calcarenite	15 – 100	Salt water aquifer
UE4	12 - > 350	Massive limestone	?	Sea water

Geoelectrical prospecting allowed integrating the data with the conceptual model and determining that the non affected area in the subsoil was located in the first 20 m. With this information, it was possible to design a drilling, sampling, and monitoring program focusing in detail to evaluate this stratigraphic position to determine the fresh water volume available and exploitable for human use.

3.1.3. Hydrogeology

The fundamental base in monitoring programs design is without doubt the previous evaluation of hydrogeological conditions of the site. Once the regional and local geology is understood, all efforts should be focused to determine at least the following: (1) local hydrogeological features of the site, (2) potentiometric surface (both at rainfall and dry seasons), (3) estimation (and preferentially the spatial variations) of hydraulic and transport parameters, and (4) geochemical and biological properties of the aquifer.

(1) Local Hydrogeological Features

As a first approach, related information of the aquifer must be gathered and analyzed to determine the geological unit or units that form the aquifer, its principal use, the rates of extraction, preferential regional flow directions, recharge, transit and discharge zones.

Once evaluated the regional behavior of groundwater, the local environment must be analyzed, establishing key points that will give vital information to fix locations and design features for monitoring wells and piezometers. At least the following must be established: (A) dominant hydrogeological system (singular homogeneous aquifer, singular heterogeneous aquifer, multiple aquifer, low hydraulic conductivity aquifer, perched groundwater that may or may not follow the regional flow pattern or secondary hydraulic conductivity aquifer), (B) hydraulic connectivity between hydrogeological units, (C) type of hydraulic connectivity between groundwater and superficial water with special emphasis at the discharge zone through rivers, streams and springs, (D) use and relevance of groundwater in the area, established from an inventory of field points that allows in a suggested 5 km radius from the affectation zone, the location of water points (preferentially active wells), depth, extraction flow, piezometric data, legal concerns, and the intended use for each catchment. Particularly in Mexico, the National Water Commission (CONAGUA) has a national database (free of access) called Public Registry of Water Rights (REPDA) that may help as an excellent starting point for a later field verification to determine all of the above mentioned parameters.

(2) Local Potentiometric Surface

Location, geometry and depth to the water table are one of the more relevant parameters of the aquifer. Even if exploratory drills are nonexistent at the site, the depth to the saturated zone could be estimated through geophysical data interpretation (see section 3.1.2) and the results of the water points inventory. This element is usually overseen in environmental studies, but when carried out adequately it can add valuable information that can hardly be obtained at such low operational costs against those of a drilling campaign. Thus it is indispensable to have hydrographs, lithologic section, constructive data, geophysical data (such as well logs), water quality data and main use for each well.

Integrated data allows designing a drilling program and the installation of piezometers to complement the missing information to configure the potentiometric groundwater flow, comparing both rainfall and dry seasons. Additionally, the exploratory drills can be used to: (a) collect soil samplings of the vadose zone to determine analytical concentrations of contaminants and for soils mechanics assessment (granulometric analysis, porosity, volumetric water content, permeability and organic matter presence), (b) determine the local lithostratigraphy and thickness of hydrogeological units, (C) groundwater sampling, and (d) perform aquifer tests to estimate the hydrogeological parameters.

The potentiometric surface configuration has the main purpose of determining the direction of the advective migration of the dissolved phase plume through the flow paths and it will serve to zone hydraulically the study area (for example, up-gradient or down-gradient zones) to be able to locate the best sites for monitoring. Also, the flow net will serve to estimate the spatial variation of the hydraulic gradient and establish interconnectivity between hydrogeologycal units or hydraulic relationships between ground and superficial water as well.

(3) Hydraulic and Transport Parameters

In classic hydrogeology, hydraulic parameters are related to the capacity of the aquifer to transmit, store, and give away water. These capacities are measurable mainly through two parameters: the transmissivity (T) and storativity (S). The estimation of hydraulic parameters represents the starting point for any serious hydrogeological analysis in regards to human use supply systems design.

In environmental studies of Contaminant Hydrogeology, more so than estimating T and S it is fundamental to determine the groundwater velocity and the velocities of the contaminants dissolved in it in order to determine a monitoring system (Barcelona, et al., 1985). The flow velocity estimation is based on the previous determination of hydraulic conductivity of the aquifer and the hydraulic gradient from the potentiometric surface according to Darcy's Law:

$$V_X = \frac{k}{\varphi_{\mathrm{ef}}} \cdot \frac{\mathrm{dh}}{\mathrm{dl}} \tag{1}$$

where:

V_x: Advective velocity of groundwater flow [LT^{-1}]
k: hydraulic conductivity of the aquifer [LT^{-1}]
Φ_{ef}: effective porosity [Adim]
dh/dl: hydraulic gradient (i) [Adim; LL^{-1}]

The hydraulic conductivity (k) can be determined by both lab and field methods, but lab estimations can vary in significant magnitude orders therefore field methods such as pumping tests, slug and tracer tests are more reliable since they yield more representative values.

Effective porosity has an influence on the capacity of geological materials to transmit and give up water through interconnected pores. There are plenty of methods to estimate effective porosity of geological materials, for example, from the interpretation of pumping tests with observation wells in unconfined aquifers (matching the specific yield with the effective porosity) or from drainable porosity and total porosity for geological materials of known granulometric conditions, as shown in figure 3.

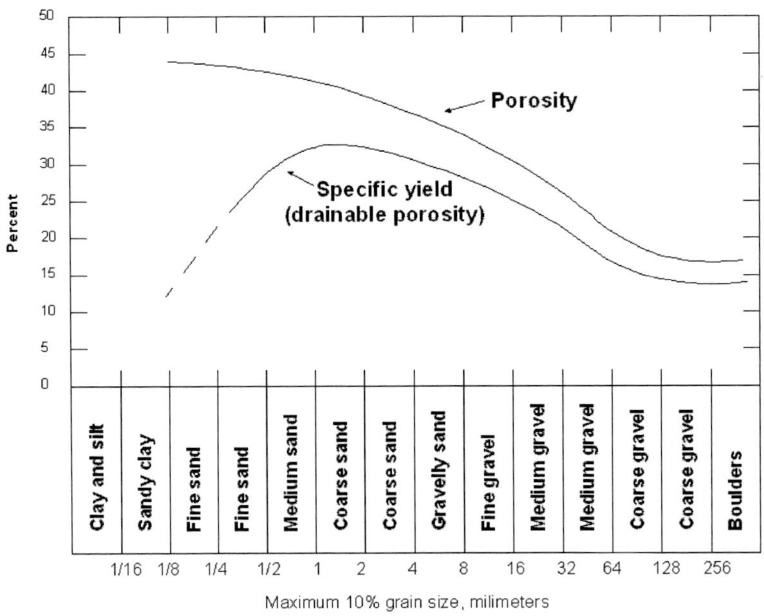

Figure 3. Total and drainable porosity for different geological materials.(Modified from Barcelona, et al., 1985).

Knowing the velocity rates of groundwater is essential to determine the location of monitoring wells at reasonable distances from the contaminant source along the flow path. Additionally to the flow velocity it is essential to determine residence times, which are mainly used in calculating sampling frequencies with special interest on monitoring potential migrations of intermittent contaminant sources or well to estimate the arrival times of contaminants to different external receivers. The equation used to calculate transit times is:

$$t_t = \frac{L_{\mathrm{AB}} \cdot \varphi_{\mathrm{ef}}}{\mathrm{ki}} \quad (2)$$

where:

L_{AB} : distance from point A (source) to point B (external receiver) along the groundwater flowpath [L].

Equations (1) and (2) are valid for groundwater advective velocity calculations (when dispersion flow does not occur) or for conservative solutes. More precise transit times analysis require nonconservative solutes from the retardation factor, so that:

$$V_C = \frac{V_X}{R_f} \quad (3)$$

where:

V_c: contaminant velocity [LT^{-1}]
R_f: retardation factor [adim].

Defined as:

$$R_f = 1 + \rho_S \left(\frac{k_d}{\varphi_T} \right) \quad (4)$$

where:

ρ_s: density of the solid phase of the aquifer [ML^{-3}]
Φ_T: total porosity of the saturated zone
K_d: distribution coefficient [L^3M^{-1}]; parameter preferentially determined in lab that depends on the lithological local features as well as those from the solute. Indicates the relationship between the solute mass incorporated to the solid phase and the one remaining as a solution, according to the linear sorption model of Freundlich (Martínez-Alfaro, et al., 2006).

(4) Geochemical and Biological Properties

Finally, the geochemical and biological properties of the aquifer matrix should be evaluated in terms of the potential interference with the goals of the monitoring program. For example, chemical reactions or biological transformations of target chemicals species may introduce artifacts into the results. Physical and hydrologic conditions will determine whether or not evidence of chemical or biological interactions can be collected. If the potential for these reactions or transformations exists, consideration should be given to screening for likely intermediates or transformation products. (Barcelona, et al., 1985).

3.2. Distribution of the Monitoring Network and Sampling Frequency

An integral part of the monitoring program is to determine the distribution and location of the monitoring wells as well as sampling frequency to evaluate the temporal behavior of the plume(s). If monitoring is considered as a primordial phase of a remediation project (in fact, that must be the case), it will be needed to design a program that allows to determine the state of the plume (stable, unstable, reduction), establish the tendency of concentrations vs time, and establish the potential migration to external receivers.

Network distribution and sampling frequency must be design according to technical criteria in the first place and secondly on budget. The guidelines of the design must be based upon groundwater displacement, the physical and geochemical properties of the aquifer and the biogeochemistry of the contaminants (Nyer, 1988). Generally, the methodology to determine both the network distribution and the sampling frequency can be divided into: (1) qualitative (based on the empirical experience of the geologist/hydrogeologist), (2) statistical (application of geostatistical and stochastic methods for optimal distribution of monitoring points), (3) analytical (application of solutions to the Advection-Diffusion Transport Equation) and (4) a combination of all the aforementioned.

3.2.1. Sampling Frequency

The sampling frequency is a very important aspect. If the samples are not collected frequently enough temporal information could be lost, but too many sampling events may generate redundant information at an excessive cost. Traditionally in Mexico, the frequency of sampling events is determined by the Environmental Authority Agency and occasionally it is estimated through statistical methods. Very often in Mexico a monthly sampling frequency is determined without taking into consideration the most important aspects.

The experience of the authors of this article, dictates that sampling frequencies for environmental projects in Mexico are overestimated, therefore it is convenient to bring up criterion to diminish those frequencies and assure the same quality during monitoring. It is convenient to consider the criterion established by Nyer (1988):

- **Frequency of sampling events:** analyze the validity of short term sampling that increase enormously the operational costs. In Mexico, it is common for environmental regulators to fix sampling on monthly basis, but depending of the site characteristics and the duration of monitoring (short, mid or long term), monthly events could make a mid or long term monitoring program money wise unbearable. When there are ground bases to considerate the advective velocity rate to be very low, high organic content in soil and biodegradable compounds with high retardations it is reasonable to think that a 6 month frequency sampling or even yearly frequency sampling would be fair enough.
- **Wind down the sampling program:** the frequency of sampling events does not have to be constant during the whole project. It is valid to reduce it as the project progresses, this is know as wind-down. For example, after 3 quarterly sampling events it is possible to reduce the frequency to a sample every 6 months. Let's say the site has been sampled quarterly for the past two years and concentrations rise somewhat during June sampling event both years. During the other six events,

concentrations remain relatively constant or drop. The Project Manager should consider reducing sampling frequency to once or twice per year, making sure to include the June event.

- **Reduce the number of wells being sampled:** for sites that have been established that the plume is stable, a review of the number of monitoring wells should be performed. There are several reasons to eliminate a particular well from a sampling event or from all future sampling events. Wells located away from the potential pathways leading to a receptor, wells located in close proximity to each other and screened in the same interval, wells screened over multiple aquifers, and poorly constructed or damage wells should all be considered for elimination. There is no reason to sample 20 wells every month just because the wells exist and they were sampled last time.
- **Reduce the analysis program:** the sampling program does not have to always include nonessential compounds in the analysis program. Secondly, the number of analytes deemed "necessary" at the beginning of the monitoring program may change over time. When a Chemical of Concern (CoC) repeatedly reaches an agreed upon threshold level or not-detect, stop analyzing it. This is especially useful when one of the compounds is not part of the main family of contaminants.

As a complement to the above suggestions it is useful to carry out a quantitative analysis to translate advective velocity into sampling frequencies according to the hydrogeological conditions of each site. Barcelona et al. (1985) designed a nomograph of sampling frequencies (figure 4) based on equation (2); rewriting this equation as:

$$F = \frac{D \cdot \varphi_{ef}}{864 \cdot ki} \tag{5}$$

where:

F: Frequency of sampling (days)
D: Distance along the flowpath (meters)

Notice how, from figure 4, the sampling frequency that must be applied to determine the migration of the plume to a potential receiver of 40 cm away along the flow path in an aquifer with hydraulic conductivity of 10^{-1} cm/s or 86.4 m/d (medium sand), hydraulic gradient of 10^{-4} m/m and effective porosity of 30 % should be 14 days, but to determine the migration to a receiver located 10 m away under the same hydraulic features the sampling frequency required would be almost yearly (347 days to be exact).

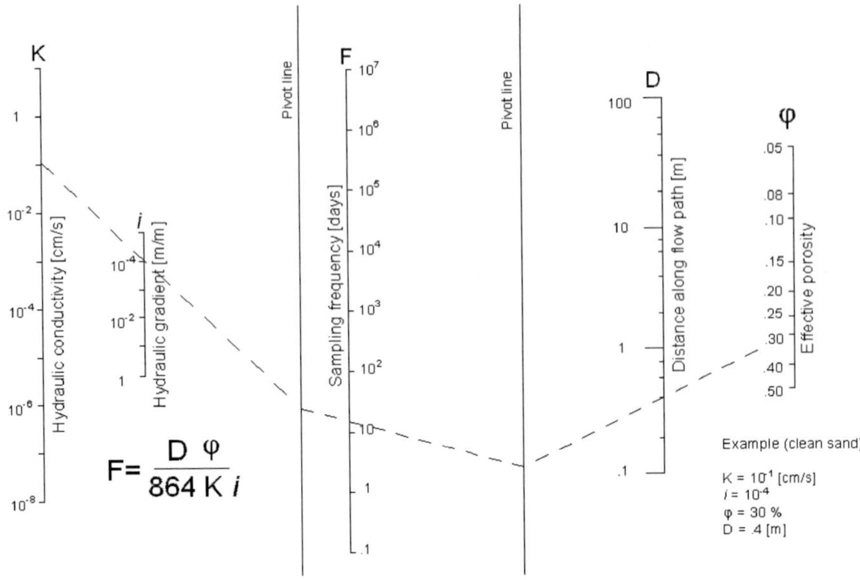

Figure 4. Sampling frequency nomograph. Modified from (Barcelona, et al., 1985).

3.2.2. Spatial Distribution of the Sampling Network

Perhaps the most important and complicated consideration during the design and implementation of a monitoring program is to determine the number and specific location of the monitoring wells and/or piezometers. First, the criteria to retain, add, or eliminate an existing well must be established, for example, when these wells where installed to carry out an environmental assessment or when an optimization of an already implemented monitoring program is required (this criterion will be further discussed). Second, it is needed to discuss the criteria needed to determine the minimal number of wells and its location when the implementation of a mid-long term monitoring program is intended.

The number of needed wells must be determined particularly for each site, and it is often convenient to do it so applying statistical methods. Resource Conservation and Recovery Act (RCRA) regulations -40CRF240- from the U.S. EPA, describe that for contaminated groundwater monitoring programs must at least consist of the following: (A) a monitoring well up-gradient (in the direction of the increase of hydraulic head at static conditions) for collecting representative samples of the background composition of the aquifer and (B) at least three monitoring wells down-gradient (in the direction of the decrease of hydraulic head at static conditions). In both cases, the monitoring wells must be screened at the same hydrostratigraphic unit to assure the data compatibility.

The RCRA criterion establishes that at least 4 monitoring wells must be installed and for its accurate location it is needed to previously characterize the local hydrostratigraphy and the configuration of the potentiometric surface and flow net. It is fundamental to remember that an erroneous interpretation of the potentiometric configuration of a contaminated site may lead to mistakes of awful consequences.

Next is shown data of a contaminated site to illustrate the relevance of the correct interpretation of a potentiometric configuration (Case-Study 2). An Environmental

Assessment Phase II was performed in a contaminated site by a diesel spill generated from a leak of a pipeline that transported hydrocarbons. The leak contaminated a volcanic sequence of unconsolidated tuffs and volcanic ashes that formed a shallow unconfined aquifer for agricultural use. The assessment consisted in the horizontal and vertical delineation of Total Petroleum Hydrocarbons (TPH) medium fraction plume in the vadose zone, free phase (LNAPL) plume, and dissolved phase of naphthalene and BTEX plume. Particularly, the LNAPL presence in the capillary zone causes interpretation problems of the aquifer's flow net, when the needed corrections are not made.

When a well penetrates a saturated stratum with LNAPL, hydrocarbons and water migrate to the interior of the well to reach equilibrium in regards to atmospheric conditions. As a result, a hydrocarbon layer is formed on the upper part of the water-table. The top of this hydrocarbon layer is known as the air-hydrocarbon interface and the bottom as water-hydrocarbon interface (API, 2004). Therefore, when a monitoring well penetrates a saturated stratum with LNAPL, the water-table depth measured inside the well is affected notoriously due to LNAPL presence (see figure 5). Actually, the measurement equals to the water-hydrocarbon measurment, therefore it is needed to transform this into the theoretical depth that the water-table should have without LNAPL presence. This transformation is done using the following equation:

$$Z_{TWLD} = Z_{OW} + (\rho_O H_O) \ldots (6),$$

where:

Z_{TWLD}: Theoretical Water Level Depth [L]
Z_{OW}: Oil-Water Interface [L]
ρ_O: LNAPL density [ML^{-3}]
H_o: Measured LNAPL thickness in the monitoring well

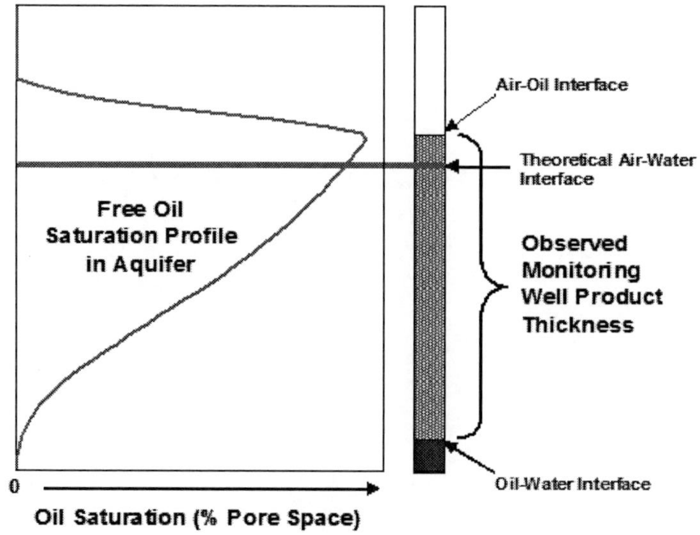

Figure 5. Conceptualization of LNAPL in a monitoring well. Modified from (API, 2004).

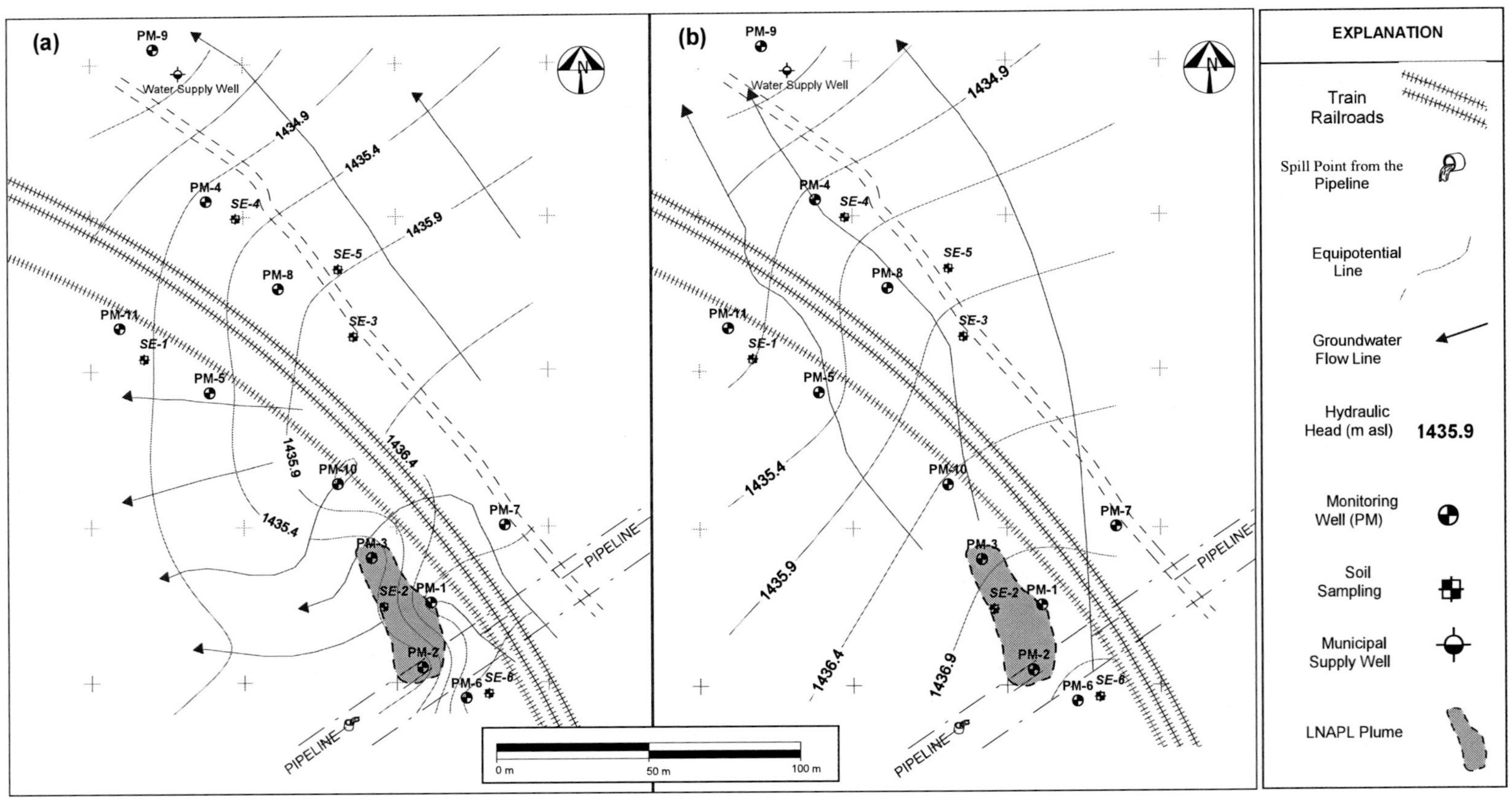

Figure 6. Distortion effect of potentiometric surface due to LNAPL presence.

In figure 6 the potentiometric configuration of the affected site is shown under two different interpretations. In the left figure (figure 6a) no consideration was taken in obtaining the theoretical water level depth, and only the direct measurement values were used to configure the flow net. For the right figure (figure 6b) the measurements were corrected using equation (6). In both cases notable differences are observed. The potentiometric surface of the left side is clearly distortioned by the LNAPL presence. Using the RCRA criteria, one up-gradient well from the LNAPL should be installed but also two well batteries in two different places. One of the batteries would be required down-gradient in the west side of the site, and the second one should be placed down-gradient northeast of the site. Once the potentiometric surface is configured with the corrected values it is clear that only one down-gradient zone should be monitored, having as a potential external receiver the municipal supply well

SPATIAL DISTRIBUTION APPLYING ANALYTICAL AND NUMERICAL METHODS

One of the most robust criteria to locate sites where monitoring point are pretended consists in applying methods based on analytical solutions and numeric approximations of contaminant dissolved phase plumes simulations under varying scenarios. There are multiple codes of numeric flow and transport modeling that allows simulating the configuration of a plume under a steady-state conditions, as well as the potential migration of the plume under a transient-state conditions with certain boundary conditions. There is also freeware such as "Optimal Well Locator" (developed by the U. S. EPA) that allows to simulate the flow and transport conditions of the evaluated site as well as displaying certain indicators that show the optimal distribution of the monitoring network so that located wells may intercept the plume in its various phases of migration (the software may be downloaded from www.epa.gov/ada/csmos/models.html).

Analytical criterion allows quantifying the contaminant migration and can be used to dimension the theoretical size of a dissolved phase plume by applying solutions for the Advection-Diffusion Transport Equation. Fetter (1999) established a methodology to determine the theoretical size of a contaminant plume which can be applied to establish potential locations of monitoring wells. The methodology states the calculation of the size of a plume transporting through time in a bidimensional flow field, as a conservative solute in a saturated porous media. The concept is based on the diffusional hydrodynamic dispersion model that predicts concentration curves to have a Gaussian distribution through time, described by its mean and variance so that:

$$D_L = \frac{\sigma^{2}{}_{L}}{2t} \; ; \; D_T = \frac{\sigma^{2}{}_{T}}{2t} \tag{7}$$

where:

D_L: Longitudinal dispersion $[L^2T^{-1}]$
D_T: Transversal dispersion $[L^2T^{-1}]$
$\sigma^2{}_L$: Variance of the longitudinal spreading of the plume [L]

σ^2_T: Variance of the transversal spreading of the plume [L]
σ: Standard deviation = $[\text{variance}]^{1/2}$
t: Time

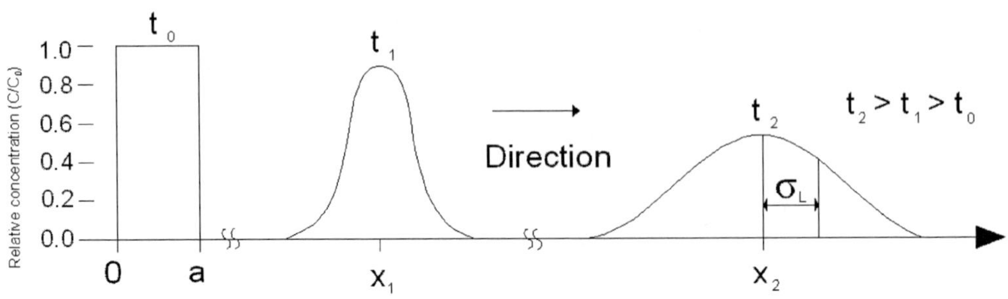

Figure 7. Transport and spreading of a solute slug with time due to advection and dispersion. (From Fetter, 1999).

The model suggests a Gaussian distribution for the contaminant in the plume with standard deviations, according to equation (7), of:

$$\sigma_x = \sqrt{2D_L t}\ ;\ \sigma_y = \sqrt{2D_T t} \tag{8}$$

In this sense, the contaminated zone will be defined by: (1) the center of mass of the plume (zone of maximum concentration), (2) the standard deviation in a vector parallel to the groundwater flow direction ("X"), and (3) the standard deviation in a vector perpendicular to the groundwater flow direction ("Y). If the groundwater flows in the "X" direction and the spill is placed at $(x_0, y_0) = (0,0)$, the center of mass of a conservative solute will be located at $(v_x t, y)$, for a time "t", v_x being the advective velocity from equation (1).

By definition, 99.7 % of a contaminant mass will be concentrated in an area represented by the position of the center of mass, three times the standard deviation in the "X" direction from the center of mass and three times the standard deviation in the "Y" direction from the center of mass (figure 8). Therefore, the plume might be defined from analytical solutions of the General Transport Equation by the position of the center of mass, $3\sigma_x$ and $3\sigma_y$ (Fetter, 1999).

This method may be applied knowing the hydraulic and transport parameters of the aquifer and may serve as a solid start point to establish the monitoring network distribution as a function of time.

3.3. Sampling Type Selection

Many are the texts that explain in detail the advantages and limitations of each type of sampling as a function of environment and contaminant conditions, as well as the strategies of development and purging and the analytical considerations surrounding quality control systems to assure reliable data (Barcelona, et al., 1985; Puls and Barcelona, 1996; USEPA, 1999; Seeve, et al., 2000; Shapiro, 2002; McCall, 2005; USDA, 2007).

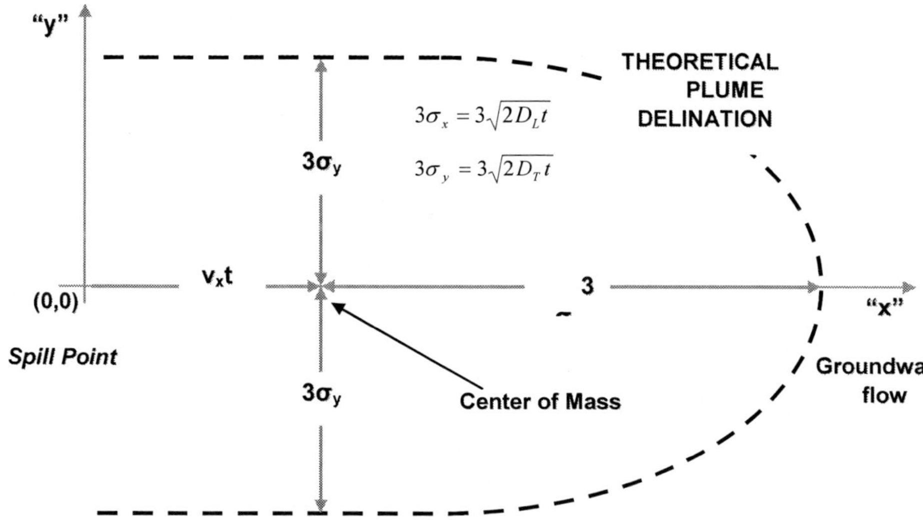

Figure 8. Theoretical plume dimensioning, supposing a conservative solute on a bidimensional flow field through saturated porous media.

This section does not pretend to re-write the previously discussed concepts; more so, it tries to show the variations that analytical results may present when different sampling methods are applied to a site in particular. Next are shown results derived from groundwater sampling campaigns in a contaminated site on the west side of Jalisco (Case-Study 3).On the site an Environmental Assessment Phase II was carried out that consisted in the hydrogeological evaluation of a mixed dissolved plume (organic volatile compounds and heavy metals) that contaminated groundwater stored in moderate to low permeability medium formed of sandy-shale deposits interspersed with volcanic deposits from Tertiary and Quaternary.

During the assessment eight monitoring wells were drilled and installed, through which two sampling events were carried out and the following parameters determined: dissolved oxygen (DO), Redox potential (Eh), electrical conductivity (EC), turbidity (Tu), temperature, benzene, arsenic, total chromium, and zinc. The first event was done using bailer and two weeks later for the second one a Low Flow Sampling method was used, both under the same geological, hydrogeological, and hydrometeorological conditions. For both cases the same multi-parametric sound was used.

As it can be seen in table 3, the results between both events have significant differences. The concentrations of the Chemicals of Concern (CoCs) obtained during the second event (Low Flow Sampling) are lower than those obtained from the first event (Bailer). The differences between concentrations for both methods range from 28.6% to 100%.

Additionally, the contaminant indicator parameters show the more significant differences. Turbidity was considerably lower in all wells when Low Flow method was used. Since turbidity measures suspended dissolved solids (SDS), a sample with a high turbidity may generate a high concentration for metals, therefore, it is coherent to think that with the Low Flow method turbidity is minimized and thus lower concentrations of dissolved metals are detected as it happens for this case. In regards to contaminant indicator parameters the most sensitive to sampling type is DO. Particularly for the site, it was crucial to determine the aerobic or anaerobic character of the aquifer, as well as the equilibrium state of the plume because depending on both the remediation strategy changed radically. The best tool to determine the aerobic or anaerobic character was in-situ DO measurement. According to table 3, DO measurements from all wells are considerably lower with the Low Flow method than those of the Bailer. Variations of DO measurements between both methods fluctuated from 54.5% to 100%, with Bailer samples the aquifer was thought to be oxygenated (2.18 to 5.49 mg/L), therefore it could be thought using remediation techniques that catalyze oxygenation such as AirSparging or simply a natural attenuation with aerobic biodegradation. Contradictorily, Low Flow sampling showed clearly that the aquifer has an anaerobic character (DO varying from 0.1 to 0.55 with a peak of 2.5 mg/L at MW-5). With this information the remediation techniques to be used are completely different focused on accelerating anaerobial biodegradation, for which specific studies were needed to determine if the aquifer had the capacity to degrade benzene in absence of oxygen through potential electron acceptors that allow the anaerobial biodegradation to catalyze the growth of microorganisms, for example NO_3^- or Fe (III) (Tiehm and Schulze, 2003). Of course, in both cases alternative dissolved metal treatment systems were needed.

The comparison of both sampling methods allowed establishing that Low Flow results are more reliable, but the use of various sampling methods helped to complement the assessment of the site and to understand better the hydrogeological conceptual model.

3.4. Automatized Aquifer Monitoring

When it has been decided that a long term monitoring program is to be implemented, the actual tendency dictates to develop an automatized monitoring system that allows collecting a great amount of temporal data while reducing costs.

It is important to mention that aquifer automatization or instrumentation does not substitute in any way the aforementioned aspects in this document; neither suggests the elimination of monitoring activities parallel to the network automatization.

Automatized monitoring suggests the implementation of instruments in the interior of a well to register temporal data regarding to the parameters that the instrument measures. The instruments introduced to the well are called ceramic transductors, devices that convert a mechanic movement generated by pressure forces (for instance, oscillations of the water-table) into an electrical signal that is used to store information into it. In the market, there is an enormous offer for this technology, but it is extremely expensive to find transductors that allow the measurement of more than five parameters. The most common and cost-effective allow the automatized measurement, storage, and transference of temporal data of hydraulic head (measured as a pressure in meters of water column), temperature, and electrical conductivity. Therefore it is imperative to carry out parallel activities to account for the measurement of other desirable parameters that are not readily available logged by transductors. It is important to notice that the most advanced technology can send data through remote or satellite signals directly into the hydrogeologist office, practically in real time, at relative accessible costs.

Table 3. Results comparison applying two different groundwater sampling methods.

PIP	MW-1			MW-2			MW-3			MW-4			MW-5			MW-6			MW-7			MW-8			PRG's (mg/L)
	BS	LFS	% D	BS	LFS	% D	BS	LFS	% D	BS	LFS	% D	BS	LFS	% D	BS	LFS	% D	BS	LFS	% D	BS	LFS	% D	
DO (mg/L)	2.18	0.3	86.24	4.1	0.42	89.8	2.3	0.55	76.1	2.61	0.1	100	5.49	2.5	54.5	3.6	0.3	91.7	3.31	0.22	93.35	4.03	1.29	68	NL
Eh (mV)	-327	-382	16.82	-250	-304	17.8	-315	-343	8.16	-336	-370	10.1	-88	-194	120	-88	-243	176	-313	-382	22.04	-33	-88	166.7	NL
pH	6.83	7.13	4.39	7.08	6.96	1.72	7.09	7.06	0.42	7.35	7.15	2.72	7.43	7.32	1.48	6.8	7.4	8.82	7.07	7.12	0.70	7.39	7.48	1.20	NL
EC (µS/cm)	7320	7540	3.01	4060	3510	13.5	8000	7910	1.13	3650	3500	4.11	3030	2750	9.24	5080	2550	49.8	6510	7140	8.82	1910	1890	1.06	NL
Tu (NTU)	33	29	12.12	64.1	10	84.4	195	16.8	91.4	52	49.3	5.19	230	68	70.4	81.6	42	48.5	222	47	78.83	40.5	23.7	70.89	NL
Te (°C)	25.7	27	5.06	23.4	23.9	2.09	23	23.6	2.54	22.7	20.5	9.69	24.9	22.8	8.43	25	23.3	6.8	22.8	23	0.87	21.8	21.6	0.93	NL
CoC's (mg/L)																									
Benzene	0.06	0.02	65.16	0.007	0.005	28.6	0.054	0.06	1.82	0.007	ND	100	ND	ND	0	ND	ND	0	0.03	0.07	57.14	ND	ND	0	0.00035
As	0.02	0.007	61.36	0.006	ND	100	0.058	0.05	13.8	0.02	ND	100	0.26	0.28	7.14	0.05	0.03	50	0.05	0.02	60	0.06	0.06	0	NL
Total Cr	0.01	ND	100	0.024	ND	100	ND	ND	0	0.04	ND	100	ND	ND	0	0.06	ND	100	0.06	ND	100	ND	ND	0	NL
Zn	0.07	0.18	59.44	0.065	ND	100	0.056	0.31	81.9	0.032	0.12	73.3	0.087	0.05	41.4	0.09	0.05	44.4	0.09	0.05	80.00	0.133	0.026	80.45	11

MW: Monitoring Well; PIP: Pollution Indicator Parameter; CoC: Chemical of Concern; BS: Bailer Sampling; LFS: Low Flow Sampling; %D: Difference percentage between BS and LFS results; red cells represent concentrations above the comparision criteria (PRGs, Region 9, U. S. EPA); grey cells represent a difference D $\geq$ 50% between both events; NL: Not Listed

One of the enormous advantages of automatized monitoring relies on the estimation of temporal variations of the parameters as opposed to conventional monitoring in which temporal information is isolated. This permits to relate variations among parameters due to data density.

To exemplify the aforementioned, data from an automatized network installed in Quintana Roo (Case-Study 4) is presented. The network was installed in the well field "Zona Captación Aeropuerto" which supplies drinking water to the city of Cancún. The extraction wells receive water from a karstic aquifer of tertiary limestones.

The network was implemented to monitor the variations of groundwater quality, quantify the advance of marine intrusion and establish the relationship salinity-hydraulic head on the recharge, transit and coastal discharge zone of the aquifer. For this purpose 10 monitoring wells were drilled and installed with depths of 40 to 100 m, PVC casing of 4" diameter, screened along the whole length and protected on its top part with a metallic cover (figure 9a). In the interior of the well two transductors were installed (to register water table and electrical conductivity variations) and connected to a Datalogger CR-200, which is the device that stores the information registered by the transductors (figure 9b). The Datalogger CR-200 programming and data downloading is carried out directly on the field through a laptop with the PC200W Datalogger software.

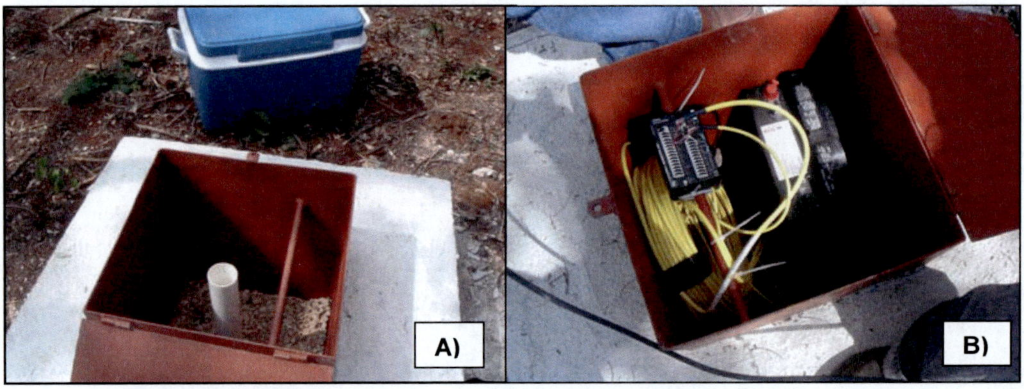

Figure 9. Monitoring network at Zona Captación Aeropuerto, Cancún. A) Metallic protection of wells. B) Monitoring instrumentation of two transductors and a Datalogger CR-200 device conected to a 12 volts marine battery.

The automatized network consisted of 10 wells that were calibrated with direct readings (salinity profiles) and programmed to record and store hydraulic head and salinity data every eight hours during seven months (as a test). The logs from April to October 2006 of PM-5, located 25 km away from the coastal border, are shown in figure 10.

The automatized logs show oscillations of the water table of 4.7 to 4.9 m and salinity variations of 2,300 to 2,900 μS/cm. Additionally, an inverse correlation between salinity and hydraulic head caused by dilution effects is clearly observed. When a certain fresh water volume incorporates to the aquifer (mainly due to rainfall water infiltration), the mixing zone moves backwards because of fresh water push and therefore an increase of the water table is observed and a diminished salinity record as there is more fresh water in the aquifer,

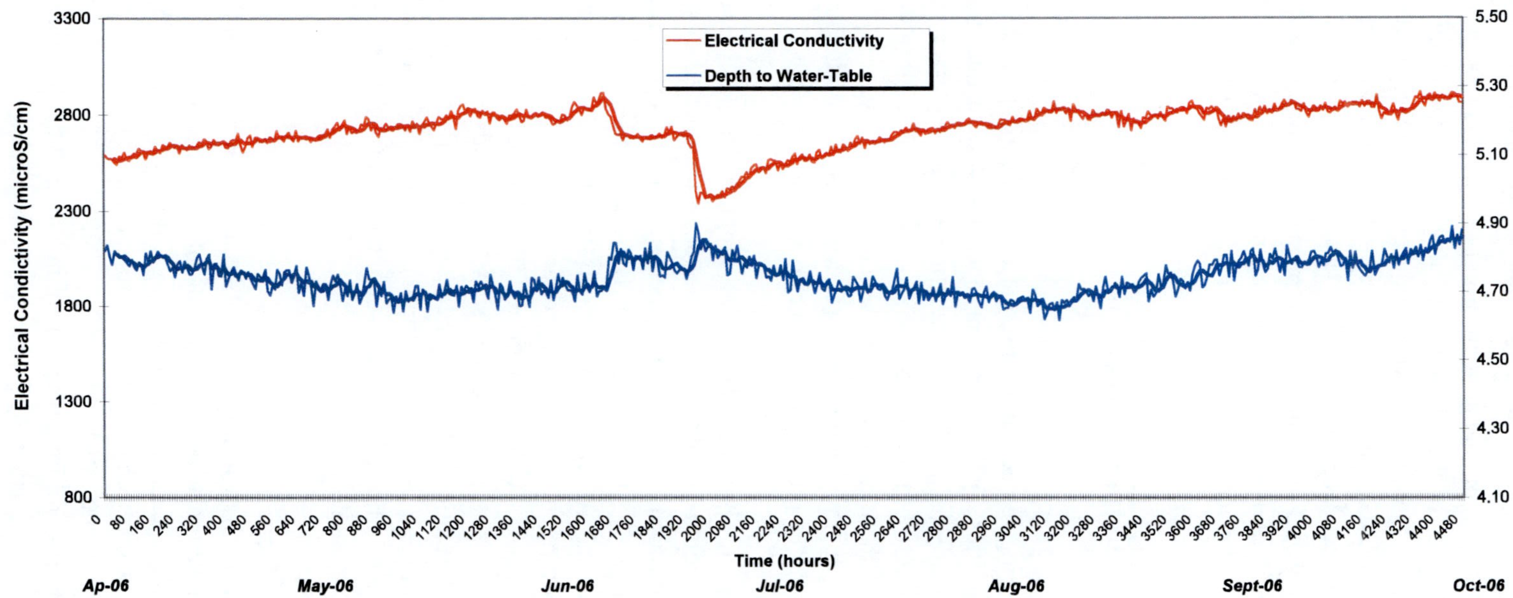

Figure 10. Results of automatized hidraulic head (blue) and electrical conductivity (red) logs from april to october 2006.

contrarily when a certain volume is drained from the aquifer (mainly due to pumping and evapotranspiration), the mixing zone is dispersed to compensate the hydrodinamics of the aquifer and the consequence is water table depletion and arise of salinity as there is less fresh water in the aquifer. This behavior is detected for the whole logged interval except on mid-September and October where a 20 cm water table level uprise produced an increase of 100 μS/cm in salinity. These continuous logs are a valuable tool to understand the behavior of the aquifer's variables that hardly could be conceptualized, if it were not by network instrumentation and automatization. Network automatization represents a highly cost-effective option for mid and long term aquifer monitoring, but again, it is important to emphasize that aquifer instrumentation must by accompanied by complementary activities.

3.5. Long Term Monitoring Programs Optimization

3.5.1. Overview

Costs for groundwater monitoring during remediation represent a significant, persistent and growing burden for the private entities and government agencies responsible for environmental remediation projects, especially as remedies are determined and implemented (USEPA, 2005). Therefore long term monitoring programs optimization positions itself more and more as an option for cost reductions without sacrificing temporal data reliability.

USEPA defines long term monitoring programs optimization (LTMPO) as the continuous re-evaluation and enhancement of monitoring programs in order to assure the same data quality while reducing operational costs.

The considerations for implementing an optimized monitoring program are many and very diverse, but at least the following should be addressed: (1) enough temporal data to establish the concentrations tendency through time, (2) the elements to be optimized (parameters, frequencies, number of wells), (3) site evaluation for LTMPO, (4) selection of optimization method, (5) implementation, (6) afterwards analysis, and (7) budget considerations. Even though the specific elements for each aspect are broadly discussed in different sources (USEPA, 2005; Asefa, et al., 2005; Aziz, et al., 2003 Dutta, 1998,) it is recommended to retake some considerations. The success of any LTMPO depends entirely of the amount and quality of the available information. At least, the following must be available (USEPA, 2005): (a) results from four to six individual sampling events (for temporal reliability), (b) results collected from six to fifteen individual sampling points (for spatial reliability), (c) a well understood conceptual model.

3.5.2. Considerations in Qualitative Evaluation

In a qualitative evaluation, the optimization regarding the numbers and locations of wells and frequency of sample collection are examined in the context of site-specific conditions. This is done to ensure that the program is capable of generating information regarding contaminant migration and changes in chemical concentrations through time and to ensure that the objectives of the monitoring programs are satisfied (USEPA, 2005).

In general, the hydrogeological conditions will influence the optimization of the locations and spacing of monitoring points, as well as the sampling frequency. Multiple factors can be considered in developing recommendations for continued monitoring, adding monitoring, or cessation of monitoring at each monitoring point/well. Typical factors considered in order to retain, add or remove a well from the monitoring network are summarized in table 4. Typical factors in developing recommendations for monitoring frequency are summarized in table 5 (USEPA, 2005).

Table 4. Qualitative monitoring network optimization decision logic.

Reasons for retaining or adding a well	Reasons for removing a well
Well is needed to further characterize the site or monitor changes in contaminant concentrations through time	Well provides spatially redundant information with a neighboring well
Well is important for defining the lateral or vertical extent of contaminants	Well has been dry for more that two years, and there is no expectation for water levels to recover in the foreseeable future
Well is needed to monitor water quality at a compliance point or receptor exposure point	Contaminant concentrations are consistently below laboratory detection limits or cleanup goals
Well is important for defining background water quality	

Table 5. Qualitative monitoring frequency optimization decision logic.

Reasons for increasing sampling frequency	Reasons for decreasing sampling frequency
Groundwater velocity is high	Groundwater velocity is low
Change in concentration would significantly alter a decision or course of action	Change in concentration would NOT significantly alter a decision or course of action
Well is close to source area or operating remedy	Well is far to source area or operating remedy
Whether concentrations will change significantly over time cannot be predicted, or there is no ready explanation for recent irregular or contradictory data	Concentrations are NOT expected to change significantly over time, or contaminant lelevs have been below cleanup objectives for some period or time

SOURCE: Roadmap to Long Term Monitoring Optimization (USEPA, 2005).

3.5.3. Considerations in Quantitative Evaluation (Temporal and Spatial Analysis)

The temporal objective of any optimization program (evaluate contaminant concentrations in groundwater through time) can be addressed by identifying trends in contaminant concentrations, by identifying periodic fluctuations in concentrations, or by estimating long term average values (Zhou, 1996), commonly applying statistical analysis. In particular tests for trends, including the Student's t-test (Zhou, 1996), regression analysis, the Mann-Kendall test (Gibbons, 1994), and (Sen's (1968) nonparametric test, are widely applied (Hirsch et al., 1991). The sampling frequency necessary to achieve the temporal objective of monitoring the can be based on trend detection, or accuracy estimation of long term average concentrations (USEPA, 2005).

Table 6. Long-Term Monitoring Optimization common tools.

LTMO Tool	Overview	Frequency Optimization Methodology	Spatial Distribution Methodology	Data Requirements	Appropriate Site Size	Typical cost to apply to site
Cost Effective Sampling (CES)	CES is a methodology for reviewing and assessing the lowest-frequency sampling schedule for a given groundwater monitoring location	Rule-based decision algorithm based on trend, variability, and magnitude statistics recommends optimal frequency at each well	Not included	- At least 6 quarterly monitoring results per well - Clean down-gradient "guard wells"	Unlimited (well-by well analysis) within same operable unit	\$25 – \$50k
Geostatistical Temporal/Spatial Optimization Algorithm (GTS)	GTS is a spatial and temporal algorithm developed by AFCEE that utilizes geostatistical methods to optimize sampling frequency and to define the network of essential sampling locations	1) Iterative thinning approach reconstructs baseline trends with fewer samples to determine optimal frequency on a well-by-well basis. 2) Temporal variogram is applied to determine composite autocorrelation and optimal site-wide frequency	Weighting scheme utilizing locally weighted quadratic regression examines multiple "time slices" to identify redundant wells based on cost-accuracy trade-off curves	- More than 8 events per well (temporal) – Greater than 30 wells (spatial)	30 to thousands of wells	~ \$25k for average site
Monitoring and Remediation Optimization System (MAROS)	The MAROS public domain software was developed in accordance with the AFCEE Long-Term Monitoring Optimization guide. The software recommends optimal future sampling frequency, location and density, as well as providing information on the plume state over time	Modified cost-effective sampling method (rule-based decision algorithm based on trend, variability, and magnitude statistics) recommends optimal frequency for each well	Weighting scheme utilizing Delaunay triangulation identifies redundant wells. Can evaluate multiple chemicals at one time	- More than 4 events per well (temporal) – Greater than 6 wells per zone (spatial)	40 to 80 wells recommended (per aquifer zone)	\$5 - \$25k for average site

SOURCE: Roadmap to Long Term Monitoring Optimization (USEPA, 2005).

Spatial techniques that can be applied to the design monitoring programs fall into several categories, such as: simulation (modeling), geostatistical (for example Kriging) and analytical approaches (such as Delaunay Triangulation) (Minsker, 2003). Once a plume model is created using one of these approaches, it can be incorporated into a numerical optimization algorithm that uses formal mathematical techniques to derive an optimal monitoring network configuration (USEPA, 2005).

3.5.4. Long Term Monitoring Program Optimization Tools

The specific optimization approach selected for a given site depends on several factors, including the amount and type of existing data, available resources and the size of the monitoring program. Current analyses tend to consider numerical simulation and formal optimization approaches, but also rule-based algorithms and professional judgment (USEPA, 2005). Table 6 lists implementation details for some of these tools.

4. CONCLUSIONS

Monitoring programs are a tool that allows to: (1) complement the environmental assessment, (2) improve the conceptual model, and (3) generate indicators to determine the success of a remediation, contention, or prevention project.

All the aspects discussed on this document point out that design and implementation of monitoring programs must be supported on sound hydrogeological bases that account for the geological features of the site, hydrodynamic particularities of the aquifer and the own contaminant properties.

It must be assimilated for the Mexican setting, the environmental relevance of groundwater, from its understanding, assessment, remediation, monitoring and protection as the key surrounding successful management of contaminated sites in Mexico.

REFERENCES

[1] Air Force Center for Environmental Excellence (2006). *Long-Term Monitoring Optimization Guide.* Environmental Restoration Division (AFCEE/TDE).

[2] American Petroleum Institute (2004). *API Interactive Guide Version 2.0.* Environmental Systems and Technologies, Division of Groundwater and Environmental Services, Inc. USA.

[3] Arango, C. (2005). *Estudio Magnetotelúrico de la zona de LLucmajor (Mallorca): Avances en el proceso de datos y modelo 3D.* PhD Dissertation, Universitat de Barcelona.

[4] Asefa, T., Kemblowski, M., Urroz, G., Mckee, M. (2005). *Support Vector Machines (SVMs) for Monitoring Network Design.* Ground Water. Vol. 43, No. 3, pages 413-422.

[5] Aziz, J.J., et al (2003). *MAROS: A Decision Support System for Optimizing Monitoring Plans.* Ground Water. Vol. 41, No. 3, pages 355-267.

[6] Barbaro, J.R. (2006). *Use of Plume Mapping Data to Estimate Chlorinated Solvent Mass Loss.* Ground Water Monitoring and Remediation. Fall 2006, pages 115-127.

[7] Barcelona, M.J., et al (1985). *Practical Guide for Ground-Water Sampling.* Illinois State Water Survey. ISWS Contract Report 374.

[8] Casas, A. (2000). *Prospección Geofísica aplicada a la Hidrogeología: Metodologías y técnicas emergentes.* In: Olmo Alarcón, M; López Geta JA (editors) *Actualidad de las Técnicas Geofísicas aplicadas en Hidrogeología.* Instituto Geológico y Minero de España (IGME), Madrid, Spain.

[9] Dutta, D., Gupta, D., Ramnarong, V. (1998). *Design and Optimization of a Ground Water Monitoring System Using GIS and Multicriteria Decision Analysis.* Ground Water Monitoring and Remediation. Winter 1998, pages 139-147.

[10] Fetter, C.W. (1999). *Contaminant Hydrogeology.* Prentice-Hall. USA.

[11] Gangopadhyay, S., et al (2001). *Evaluation of Ground Water Monitoring Network by Principal Component Analysis.* Ground Water. Vol. 39, No. 2. pages 181-191.

[12] Gibbsons, R.D. (1994). *Statistical Methods for Groundwater Monitoring.* John Wiley & Sons, Inc. New York.

[13] Hassan, A.E. (2006). *Developing a Long-Term Monitoring Network Under Uncertain Flow paths.* Ground Water. Vol. 44, No. 5, pages 710-722.

[14] Hirsch, R.M., Alexander, R.B., Smith, R.A. (1991). *Selection of Methods for the Detection and Estimation of Trends in Water Quality.* Water Resources Research Bulletin. 27(5), pages 803-813.

[15] Martínez-Alfaro, P.E., Martínez-Santos, P., Castaño-Castaño, S. (2006). *Fundamentos de Hidrogeología.* Ediciones Mundi-Prensa, Madrid, Spain.

[16] McCall, W. (2005). *Evaluation of a Small Mechanical and Pneumatic Bladder Pump for Water Quality Sampling.* Ground Water Monitoring & Remediation Vo. 25, No. 2, Spring 2005, pages 142-153.

[17] Minsker, B. (2003). *Long-Term Groundwater Monitoring-The State of Art.* American Society of Civil Engineers. Reston, VA.

[18] Murray, D.E & Cherry, J.A. (2002). *A New Multilevel Ground Water Monitoring System Using Multichannel Tubing.* Ground Water Monitoring and Remediation. Fall 2002, pages 52-65.

[19] Nyer, E.K. (1988). *Ground Water and Soil Remediation. Practical Methods and Strategies.* ARCADIS Geraghty & Miller. Ann Arbor Press.

[20] Puls, R.W., and Barcelona, M.J. (1996). *Low-Flow (Minimal Drawdown) Ground-Water Sampling Procedure. EPA/540/S-95/504*

[21] Sara, M.N. (2003). *Site Assessment and Remediation Handbook.* Second Edition. Lewis Publishers.

[22] Seeve, J.E., et al (2000). *An Analysis of Low-Flow Ground Water Sampling Methodology.* Ground Water Monitoring and Remediation. Spring 2000, pages 87-93.

[23] Sen, P.K. (1968). *Estimates of the regression coefficient based on Kendall's tau.* Journal of the American Statistical Association Vol. 63, pages 1379-1389.

[24] Shanklin, D.E., et al., (1995). *Micro-Purge Low-Flow Sampling of Uranium-Contaminated Ground Water at the Fernald Environmental Management Project.* Ground Water Monitoring and Remediation. Summer 1995, pages 168-176.

[25] Shapiro, A.M. (2002). *Cautions and Suggestions for Geochemical Sampling in Fractured Rock.* Ground Water Monitoring & Remediation 22, No. 3, Summer 2002, pages 151-164.

[26] Smith, R.T & Ritzi, W. (1993). *Designing a Nitrate Monitoring Program in a Heterogeneous, Carbonate Aquifer.* Ground Water. Vol. 31, No. 4, pages 576-584.

[27] Tiehm, A , Schulze, S. (2003). *Intrinsic Aromatic Hydrocarbon Biodegradation for Groundwater Remediation.* Oil & Gas Science and Technology. Vol. 58, No. 4. pages 449-462.

[28] U. S. EPA (2005). *Roadmap to Long-Term Monitoring Optimization.* EPA 542-R-05-2003.

[29] U. S. EPA (2004). Guidance *for Monitoring Hazardous Waste Sites – Framework for Monitoring Plan Development and Implementation.* U. S. EPA Office of Solid Waste and Emergency Response. Directive, No. 9355.4-28.

[30] U. S. EPA (1999). *Site Characterization for Subsurface Remediation.* Center for Environmental Research Information. EPA/625/4-91/026.

[31] U.S. Department of Agriculture and Forest Service (2007). *Technical Guide to Managing Ground Water Resources.*

[32] Van Blaricom, R. (1992). Practical Geophysics II for the Exploration Geologist. Northwest Mining Association. USA.

[33] Zhou, Y. (1996). *Sampling Frequency for Monitoring the actual state of groundwater systems.* Journal of Hydrology 180(3):301:31.

In: Remediation of Soils and Aquifers
Editors: Luis G. Torres and Erick R. Bandala
ISBN 978-`-60741-372-1

Chapter 9

CONSTRUCTION AND OPERATION OF BIOPILES

Rosario Iturbe Argüelles
Instituto de Ingeniería-UNAM

1. INTRODUCTION

Bioremediation is a greatly accepted topic nowadays that implies a process by which the organic contaminants present in the soil, due to anthropogenic activities, are biodegraded. The final stage of biodegradation is mineralization, the transformation of organic contaminants into CO_2, water, and biomass.

The objective of bioremediation is to maintain the maximal possible growth of microorganisms until the carbon source (organic contaminant) decreases and, in consequence, the microbial population decreases.

Physical and chemical factors, such as water content, temperature, pH, oxygen, and nutrients are among the most important for efficient bioremediation.

Water

Water content is one of the most important factors for degradation because it constitutes 80 to 90% of the weight in the molecular composition of bacterial cells and is one of the most important factors for the growth of microorganisms.

Ph

The pH value for bacteria to strive lies between 6.5 and 7.5, therefore, this is the interval at which the soil must be kept for an optimal growth of microorganisms.

Temperature

Chemical and enzymatic reactions increase in the cell with increasing temperature. Each organism has a minimal temperature, below that no growth occurs; likewise there is an optimal temperature at which growth is the fastest, and a maximal, above it, no growth occurs either. The interval for the optimal growth of most bacteria is between 20 and 35 °C (Yong and Mulligan, 2004).

Oxygen

Oxygen is the most used electron acceptor by microorganisms to degrade organic compounds. If the oxygen content of soil is below 1%, the aerobic and anaerobic conditions will change.

Aerobic Respiration

Aerobic microorganisms have enzymatic systems able to oxidize organic compounds that serve as electron donors. This is the most efficient way of a biochemical reaction by which the organic substrate is degraded, producing biomass, CO_2, and water (Boettcher and Nyer, 2001).

Nutrients

The solid portion of the cell is constituted by carbon, nitrogen, hydrogen, oxygen, phosphorus, and, in a lesser amount, by potassium, sodium, calcium, magnesium, chloride, iron, among others. The main component is carbon (50%). The contaminant to be degraded must contain this element. After carbon, oxygen occupies the most space (20%). Oxygen is required for the new cells and as electron acceptor, therefore large amounts of oxygen are needed for biological degradation.

The other major nutrients needed by the organisms are nitrogen and phosphorus. The three main forms of nitrogen in microorganisms are proteins, cell wall components, and nucleic acids. Phosphorus, in the form of inorganic phosphates, is used by the microorganisms to synthesize phospholipids and nucleic acids (Boettcher and Nyer, 2001).

During the bioremediation process, microorganisms use the organic contaminants for their growth. These compounds also supply nutrients, such as nitrogen and phosphorus, as well as minor nutrients, such as sulfur and trace elements, that are also important for growth. Heterotrophic organisms are those using organic compounds as carbon and energy source (Suthersan, 1997)

There are two types of transformations: in the first, carbon and energy supply is necessary for the growth of microorganisms; in the second, carbon is used by the cells for respiration.

Before bioremediation, there is a period called acclimation or adaptation period. The duration of this period depends on the chemical structure and concentration of the compounds and the environmental conditions.

Within the bioremediation process, there are diverse application techniques:

- *in situ* bioremediation
- biopiles,
- land farming,
- phytoremediation.

This chapter will deal particularly with the biopiles method, since this is the method that has been applied successfully by the Remediation of Soils and Aquifers Research Group from the Engineering Institute of UNAM in the state of Tabasco in 2006.

Biopile is defined as a biological process that is part of the bioremediation methods used to treat soils contaminated with hydrocarbons. The process consists of forming piles with the contamined soil and stimulates microbial activity by aerating and adding nutrients and water. The increase in microbial activity is directly proportional to the reduction in hydrocarbon (HC) concentrations.

The objective of biopiles is to reduce the concentration of hydrocarbons adsorbed in the contaminated soils, by means of biodegradation and itsefficiency depends on several parameters grouped in three main categories:

- soil characteristics,
- contaminant characteristics,
- climatic conditions.

Soil type is utmost relevant because water, nutrients, and air must be able to migrate through the soil pores in order to allow microorganisms to accomplish degradation. Texture of the soil influences permeability, water content, and soil density. Soils with high permeability are the easiest to aerate and therefore, the best suited for the use of biopiles.

Biopiles are constructed on an impermeable base to reduce leaching of contaminants to the subsoil. Likewise, a network of perforated or slotted pipes is installed inside the pile and connected to an aeration system to allow the air to penetrate the soil.

Advantages

- The contaminants are transformed into more simple compounds and can become mineralized.
- Biopiles are easy to design and construct.
- Soil restoration is accomplished in a 3- to 6-month period, depending on the climatic conditions, the hydrocarbons concentration, and the type of soil.
- The biopile-restored soil can be returned to its original site.
- Biopiles offer a more competitive cost than physicochemical techniques.

LIMITATIONS

- Polycyclic aromatic hydrocarbons (PAHs) of five and six rings are hardly degraded in biopiles.
- Biopiles are not effective for total hydrocarbon concentrations above 50 000 mg/kg.
- The presence of heavy metals can inhibit microorganisms' growth.

TECHNICAL BASES FOR THE DESIGN AND CONSTRUCTION OF A BIOPILE (FAHNESTOCK ET AL., 1998)

- Preparation of the site
- Preparation of the base
- Collection of leachates
- Aeration
- Addition of water
- Addition of nutrients

Preparation of the Site

The area must count upon adequate space and services that will allow for the installation of the aeration and water distribution systems. A flat surface is required for the construction of the biopile. The space must be large enough to store the soil, mix it, and prepare it for the biopile.

Preparation of the Base

The base of the biopile has three functions:

- To provide a stable foundation to support the biopile as well as to manage the required procedures.
- Provide a barrier against the potential migration of contaminants to the subsoil (leaching).
- Avoid leachates accumulation
- The base of the biopile can be made with a layer of compacted clay or a paved area.

Collection of Leachates

For a permanent biopiles system must be installed a system to collect the leachates. It usually consists of a gutter placed on the side of less slope of the biopile witha perforated pipe in the lower part of the pile, and a pump to extract the leachates, connected to a drainage and a storage tank. When the biopile is constructed with the optimal amount of water, it does not produce leachates (Fahnestock et al., 1998)

Aeration

The biopile must be designed with an aeration system to provide enough air, i.e., oxygen, for the microorganisms to be able to degrade the compounds. Aeration can be passive or active. The first consists of installing drainages, perforated or slotted, placed at different heights throughout the biopile so the air can penetrate naturally, without an electrical venting system.

The active ventilation systems take into account air injection and extraction systems. These systems must perform in such a way that the airflow is exactly above the limiting oxygen conditions to avoid excessive hydrocarbon volatilization and, thereby, avoiding vapor emissions to the outside. If vapors are excessive, an activated carbon system can be used. When dealing with heavy hydrocarbons, as those originating from crude or heavy oils, there is no need for a vapor extraction system.

Water Addition

There must be water in the biopile, but not in excess, since microorganisms require water for nutrients transport and to accomplish their metabolic processes and maintain cell structure. Too much water is undesirable because when water occupies a large portion of the soil pores, it limits air permeability reducing aeration efficiency. On the other side, an excess of water increases the generation of leachates from both hydrocarbons and nutrients.

Water content can change along the remediation process. Air generally penetrates the biopile at less than 100% of relative humidity. Air tends to remove moisture as it moves through the biopile and starts to saturate with water, thereby reducing water content. On the other side, biodegradation process converts the hydrocarbons into CO_2 and H_2O, returning water to its original content.

In general, water content must be maintained between 40 and 85% of the field capacity, although it is recommended that soils have a water content between 70 and 90% of the field capacity if water is going to be added only during the construction stage.

Addition of Nutrients

The microorganisms that degrade hydrocarbons in a biopile use them as carbon and energy sources. Organisms require a carbon source to produce biomass. The contaminants and the natural organic compounds of the soil generally supply the needed carbon, but there are other substances that are essential nutrients, like nitrogen, phosphorus, and potassium, which can be found at a lower proportion with respect to carbon. The U.S. EPA recommends a carbon: nitrogen: phosphorous (C:N:P) relationship of 100:10:1 or 100:10:0.5.

Nutrients are dissolved in water and sprayed on the soil or used in a granular form adding them to soil before the construction of the biopile.

2. Practical Case

In 2006, the biopile technology was applied at a site contaminated with hydrocarbons.

The site, located in the southeast of Mexico in the state of Tabasco, in the ;Maritime Terminal of Dos Bocas. In this region, 25 000 tons of soil contaminated with heavy fraction hydrocarbons (HFH), originating from the crude operations performed for more than 30 years, were stored. The average concentration of the contaminated soil was of 16 g/kg. Mexican regulations (NOM-138-SEMARNAT/SS-2003) allow for a maximum permissible level (mg/l) of 6000 mg/kg for industrial use.

Before starting the biopiles, samples were taken from the stored soil at six different points. A compound sample of the soil was made and analyzed. The sample was used to characterize the soil in the laboratories of the Instituto de Ingeniería UNAM. Polyaromatic hydrocarbons (PAHs) were determined in the sample as well. Treatability tests were also performed at laboratory level to verify the viability of cleaning the contaminated soil using biopile.

3. Material and Methods

The main characteristics of the contaminated soil were obtained and are shown in Table 1. Figure 1 depicts the granulometry curve of the soil.

Table 1. Soil characterization.

Parameter	Method	Result
pH	Potentiometer (ASTM-D 4972-89, 1993)	7.41
Water content	Oven (NMXAA-16, 1984)	1%
Bulk density	Test tube method (Domínguez and Aguilera, 1990)	$\rho = 1.2$ g/cm^3
Dry density	Pycnometer (ASTM; D854-83, 1995)	$\rho = 2.6$ g/cm^3
Porosity	From the densities	0.52
Organic matter	Walkley and Black (NOM 021-RECNAT 2000)	MO = 5.39%
Organic carbon fraction	Walkley and Black (NOM 021-RECNAT 2000)	$f_{oc} = 3.0$ %
Granulometric analysis	SUCS (ASTM, D 422)	Figure 1
Total nitrogen	Kjeldhal (DOF, 2002b)	524 mg/kg
Available phosphorus	Bray-P1 (DOF,2002b)	2.37 mg/kg
Heterotrophic bacteria counting	Plate counting (NOM 092-SSA-1994)	8.7 x 10^8 CFU/g

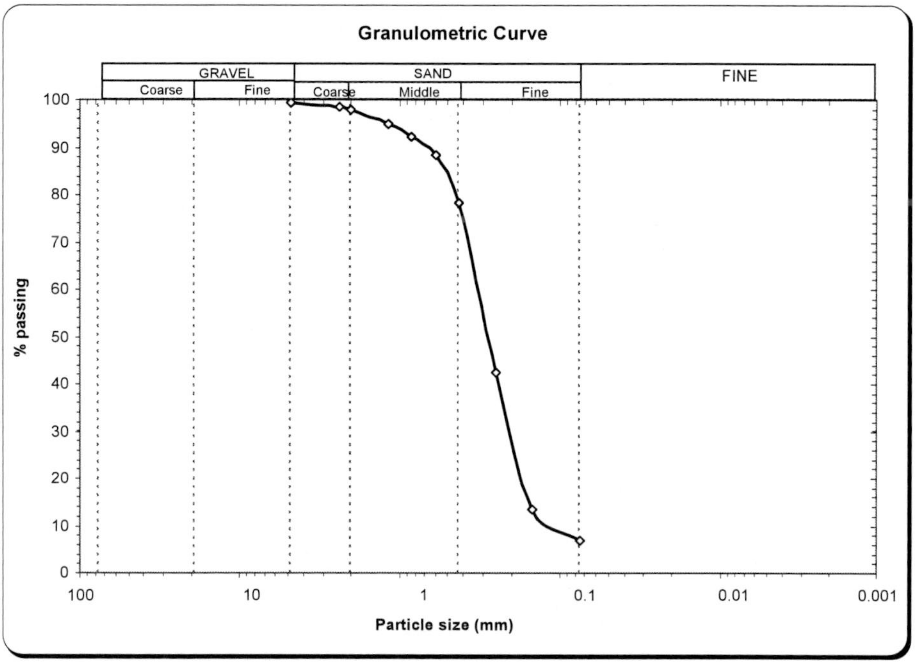

Figure 1. Granulometry curve of the studied soil.

Field work consisted of applying the biopiles on the problem site. The result of the previously performed treatability tests indicated the feasibility to apply the biopiles method, using nutrients with compressor-driven aeration.

Figure 2. Separation of residues through a cribble.

The contaminated soil was removed from the storage site by means of retro-excavators, which conveyed the soil to a coarse matter separator, designed by II-UNAM (Figure 2), aimed at separating residues like stones, wood, junk, and garbage contained in the stored soil. The soil without these residues was stored to construct the biopile.

Before start building the biopile, the terrain was leveled off and then a 1000-caliper geomembrane was placed as a liner for each biopile (Figure 3).

Figure 3. Placement of the liner for the biopile.

In September 2006, the construction of the six biopiles started. Each biopile was 60 m long, 20 m wide, and 1.80 m high, with a 1,5:1 slope.

Each biopile was constructed in layers of 0.30 m thickness. A 1"-diameter PVC perforated pipe was placed on top of each layer for aeration purposes (Figure 4).

Figure 4. PVC pipeline for aeration.

The pipeline was placed following the width of the biopile at 5 m intervals. Likewise, nutrients such as nitrogen and phosphorus, in the form of urea and triple phosphate, were applied to supply the soil with the necessary substrate for the degrading bacteria. The nutrient dose was based on the carbon, nitrogen, and phosphorus (C:N:P) relation, i.e., 100:10:1.

Figure 5. Gravel layers on top of each aeration pipe (Iturbe et al, 2008).

Figure 6. Construction of the third layer.

Table 2. presents the data used to calculate the amount of required nutrients.

Table 2. Areas and volume of each biopile layer.

Layer	Area (m^2)	Volume (m^3)	W (ton)
1	5.865	44.28	70.85
2	5.595	37.2	59.52
3	5.325	30.62	49.00
4	5.055	24.52	39.23
5	4.785	18.90	30.24
6	7.375	20.28	32.45
Σ	33.98	175.8	282.3

Nutrients were calculated from the initial concentrations of available total nitrogen (524 mg/kg) and phosphorus (2.4 mg/kg) in the contaminated soil. Accordingly, requirements for these parameters were 756 mg/kg and 125.6 mg/kg, respectively.

Before supplying the aeration system, the airflow per cubic meter of soil must be calculated.

Two kilograms of O_2 are needed for each kilogram of total hydrocarbons. Each biopile had a volume of 1784 m^3 and a density of 1.6 ton/m^3 (2855 ton of total mass) of. There were 2855 x 10^3 x 0.016 = 45 680 kg of HC; hence, 91.4 ton of O_2 were required. In 150 days the requirements were 609 kg/day = 423 g/min. To reach this amount, a 15-HP compressor was used.

AERATION SYSTEM

Figure 7. Air supply.

Once the six layers of soil had been placed, the biopiles were covered with gravel to protect them from the environmental conditions (wind and rain) (Figure 8).

Figure 8. Gravel layer on top of the biopile.

Figure 9. shows the finished biopiles after six months of treatment.

Figure 9. Set of finished biopiles.

Water content and pH were monitored every two weeks, maintaining a pH of 7.5. The water content varied between 19 and 25%. The bulk density was 1.6 ton/m^3.

Samples were taken once a month from each biopile, positioning the sampling points by means of a GPS. For the biopiles, four sampling points were located along the north side, four along the south side, and four points on the upper surface of the biopile. At each sampling point, two or three depths were considered, the first between the surface and 0.50 m, the second between 1.0 m and 1.50 m, and between 1.70 and 2.80 m depht (Fig 2).

RESULTS

Hydrocarbon removal

Biopiles B1 to B4 had four sampling points in each of their sides, they were named as follows:

- South side: SU1, SU2, SU3, and SU4
- North side: N1, N2, N3, and N4
- Upper side: S1, S2, S3, and S4

Figure 10 depicts a general diagram of the location of the sampling points for the biopiles and figure 11 shows the sampling of one of the biopiles .

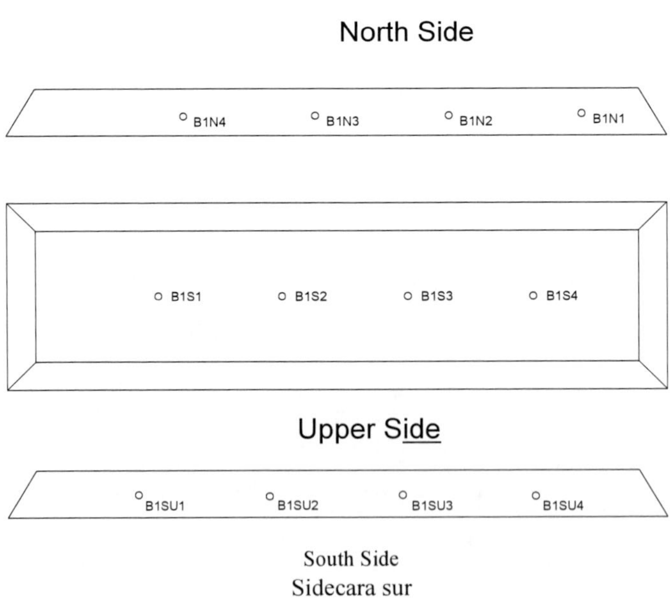

Figure 10. Sampling points diagram of the biopiles.

Figure 11. Sampling of one of the biopile.

Figure 12. Sample collection after its homogenization.

Figures 13 to 16 depict the results for heavy fraction hydrocarbons (HFH) from the samplings during the weeks that the biopiles remained active. The line indicating the limit refers to the maximum permissible value according to the NOM.138-SEMARNAT/SS-2003.

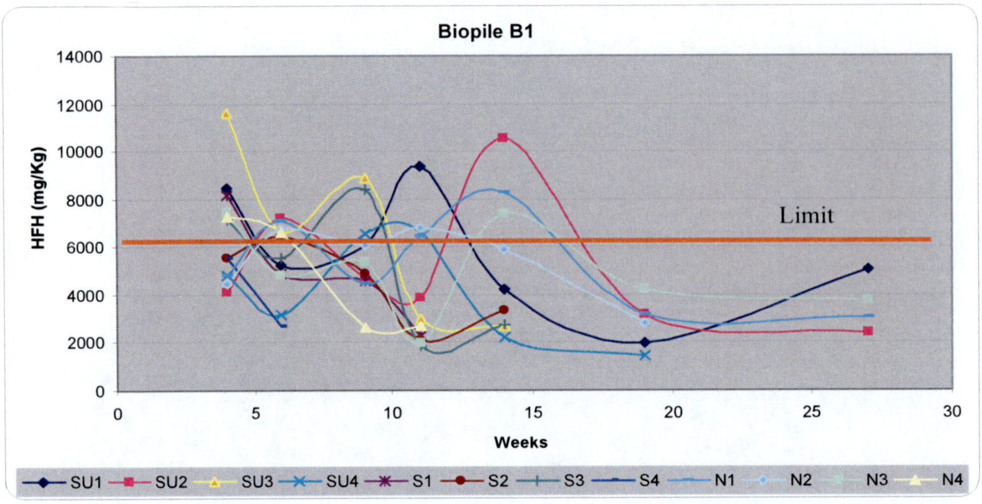

Figure 13. Variation of HFH in biopile B1.

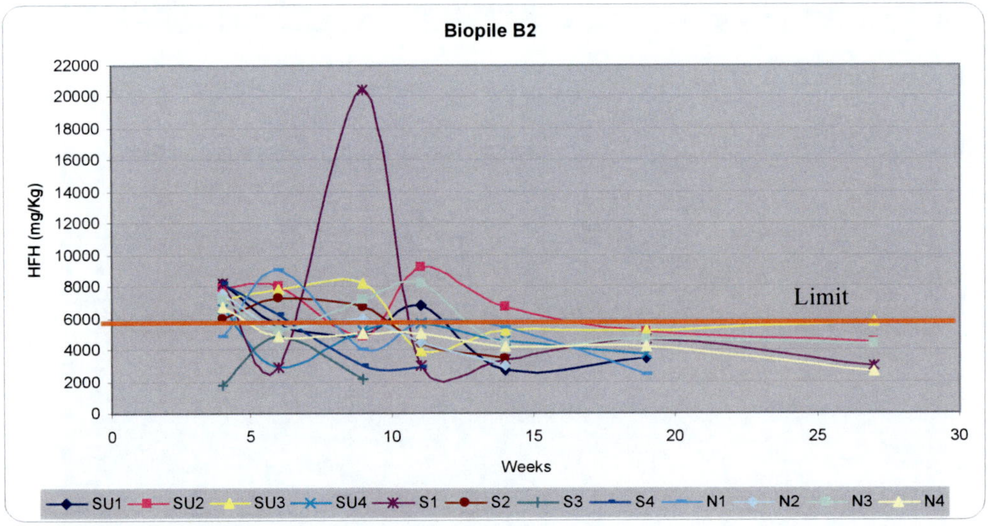

Figure 14. Variation of HFH in biopile B2.

For biopile B1, HFH values below the permissible limit start to appear at 18 weeks and for biopile B2 at approximately 17 weeks. In this case, a very high HFH value was encountered (> 20 000 mg/kg). This situation is frequent in samplings of soil with hydrocarbons, due to the presence of a solid hydrocarbon fragment at the time of sampling.

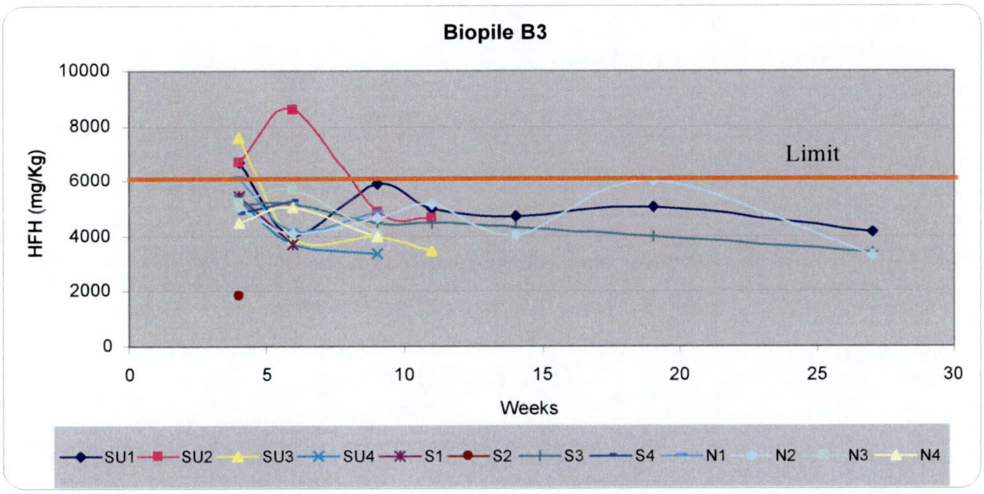

Figure 15. Variation of HFH in biopile B3.

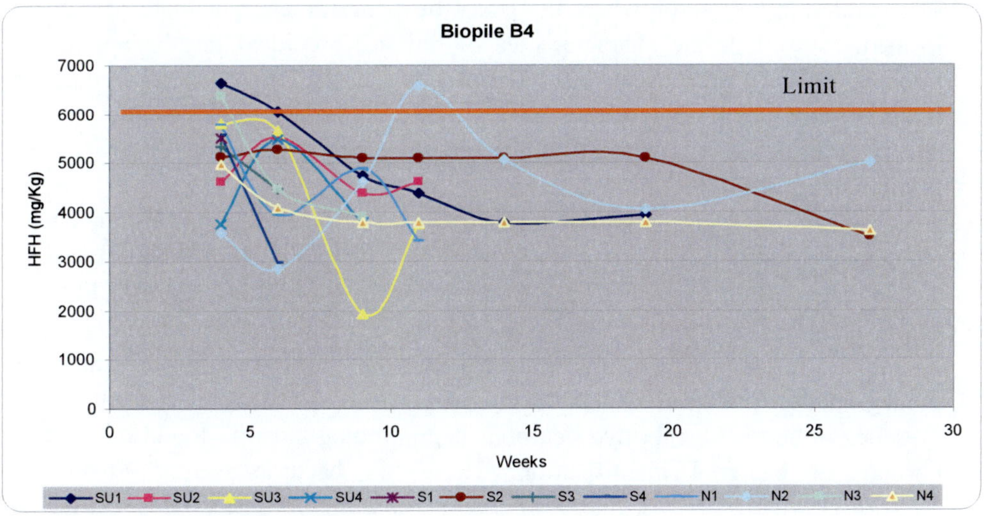

Figure 16. Variation of HFH in biopile B4.

Biopile B3 presented values below the permissible limit starting at 10 weeks after installing the biopiles, and biopile B4 at 8 weeks. This can have been due to better aeration processes in these biopiles because they had the same number of aeration pipes in a smaller area with respect to biopiles B1 and B2. The area of biopiles B3 and B4 was 60 m long and 18 m wide, instead of that used for the other biopiles (70 m length and 20 m width).

The permissible maximal limits were not exceeded for either the mid fraction hydrocarbons or PAHs in any of the cases.

None of the biopiles generated any leachates.

Bacteria identification studies confirmed the presence of six hydrocarbon-degrading bacteria in the biopile samplings. The identified bacteria were:

Flavobacterium odoratum
Bacillus brevis
Stenotrophomonas maltophilia
Staphylococcus vituluss
Staphylococcus sedentarius
Micrococcus roseus

The main conclusions reached based on the results obtained with biopiles are presented.

CONCLUSIONS

- The bioremediation method by means of biopiles applied at the Dos Bocas maritime depot resulted adequate for the removal of hydrocarbons in the 15 260 m^3 of soil. HFH concentrations were reduced from 16 000 mg/kg until reaching the maximal permissible limits established by the NOM-138-SEMARNAT/SS-2003, which is of 6000 mg/kg for industrial land use.
- The climate in the region where the Dos Bocas maritime depot, state of Tabasco, is located, as well as the characteristics of the soil are ideal for the application of biopiles. On one side, because the temperature and pH levels correspond to those indicated at the beginning of the chapter and, on the other side, the sandy soil of the site presents a texture that allows for the flow of air to be distributed uniformly through the pores of the soil.
- According to the microbiology results, biopiles from the beginning to their end had an adequate number of bacteria for hydrocarbon biodegradation (Fahnestock, 1998). The average bacterial concentration of the soil samples was of 2.3 x 10^3 CFU/g.
- There is biodegradation in the hydrocarbons removal process since the bacterial counts and respirometry results indicated that they comply with the conditions for biodegradation (Fahnestock, 1998). Hydrocarbons removal can be attributed to the presence of any of the six hydrocarbon- degradinding bacteria found in the samples chosen for bacterial identification. These six bacteria were: *Flavobacterium odoratum, Bacillus brevis, Stenotrophomonas maltophilia, Staphylococcus vituluss, Micrococcus sedentarius,* and *Micrococcus roseus.*

REFERENCES

[1] ASTM American Society of Testing Materials (1993) Method C-128-93 Standard Method for Specific Gravity and Absorption of Fine Aggregates, Vol 04.02.

[2] Atlas, R. (1983) Microbial degradation of petroleum hydrocarbons: an environmental perspective. Microbiolgical Reviews. Vol 45 (1): 180-209.

[3] Berg, G. Roskot, N. Smalla, K. (1999) Genotypic and phenotypic relationships between clinical and environmental isolates of *Stenotrophomonas maltophilia*". *J Clin Microbiol* 37 (11): 3594-600. PMID 10523559.

[4] Boettcher G. and Nyer E. K. (2001) "In Situ Treatment Technology" . Lewis Publishers. ISBN 1-56670-528-2

[5] Corona, R. (2004) Atenuación natural en suelos contaminados con hidrocarburos. Thesis. Universidad Nacional Autónoma de México. División de Estudios de Postgrado Facultad de Ingeniería.

[6] Fahnestock, F. G.B. Wickramanayake, R. Mayor, W. (1998) Biopile design, operation, and maintenance handbook for treating hydrocarbon-contaminated soils. Battelle press. USA.

[7] Iturbe, R. Flores, C. Bautista, G. Torres, L. (2004).Remediation of contaminated soil using soil washing and biopila methodologies at a field level. Soil & Sediments 4(2): 115-122.

[8] DOF (2005). NOM-138-SEMARNAT/SS-2003 Límites máximos permisibles de hidrocarburos en suelos y las especificaciones para su caracterización y remediación.

[9] Nyer, Evan., 1993. Practical Techniques for Groundwater and Soil Remediation. USA.

[10] Palleroni, N. Bradbury, J. (1993) *Stenotrophomonas*, a new bacterial genus for *Xanthomonas maltophilia* (Hugh 1980) Swings et al. 1983". International Journal of System Bacteriology 43 (3): 606-9. PMID 8347518.

[11] Stackebrandf, E. Kock, C. Gvozdiak, O. Schumann, P. (1995) Taxonomic Dissection of the Genus *Micrococcus: Kocuria gen.nov., Nesterenkonia gen.nov., and Micrococcus Bacteriology*. International Journal of Systematic Bacteriology 45 (4): 682-692.

[12] Suthersan Suthan S. (1997) "Remediation Engineering Design Concepts" Lewis Publishers. ISBN 1-56670-137-6.

[13] Webster, J. Bannerman, T. Hubner, R. Ballard, D. Cole, E. Bruce, J. Fiedler, F. Schubert, K. Kloos, W. (1994) Identification of the *Staphylococcus sciuri* species group with EcoRI fragments containing rRNA sequences and description of *Staphylococcus vitulus sp.* nov. Int J Syst Bacteriol. 44(3): 454-60.

[14] Yong R.N. and Mulligan C.N. (2004) "Natural Attenuation of Contaminants in Soils" Lewis Publishers, ISBN 1-56670-617-3.

[15] Yağcı, A. ÇerikÇioğlu, N. Kaufmann, M. Malnick, H. Söyletir, G. Babacan, F. Pitt. T. (2000) Molecular Typing of Myroides odoratimimus (Flavobacterium odoratum) Urinary Tract Infections in a Turkish Hospital. European Journal of Clinical Microbiology & Infectious Diseases 19 (9): 731-732.

In: Remediation of Soils and Aquifers
Editors: Luis G. Torres and Erick R. Bandala

ISBN 978-1-60741-372-1

Chapter 10

SURFACTANT FACILITATED REMEDIATION OF DNAPL IMPACTED SUB-SURFACE ENVIRONMENTS

Stephen Leharne
University of Greenwich at Medways. Kent, England

INTRODUCTION

Organic liquids such as petroleum hydrocarbons, industrial solvents and oils are of wide-ranging economic importance. Their prevalent use in manufacturing industries, however, provides enormous opportunities for environmental contamination. The presence of organic liquids as separate liquid phases, in the sub-surface arises because of spills, former and current disposal procedures and leakages from storage facilities. Organic liquids which have a specific gravity greater than 1 can sink below the water table, where they may be encountered as: still mobile bodies; discontinuous residual droplets trapped in the pores of sedimentary deposits and in bedrock fractures or as potentially mobile pools that have accumulated at capillary barriers. Figure 1 provides an example of coal tar droplets trapped in a synthetic porous medium comprising glass beads and originally saturated with water.

Figure 1.

Once in groundwater these organic liquids, which are frequently called dense non-aqueous phase liquids (DNAPLs) – to emphasise their immiscibility with water and specific gravity greater than 1 – slowly dissolve in flowing groundwater to create an aqueous phase contaminant plume (Figure 2).

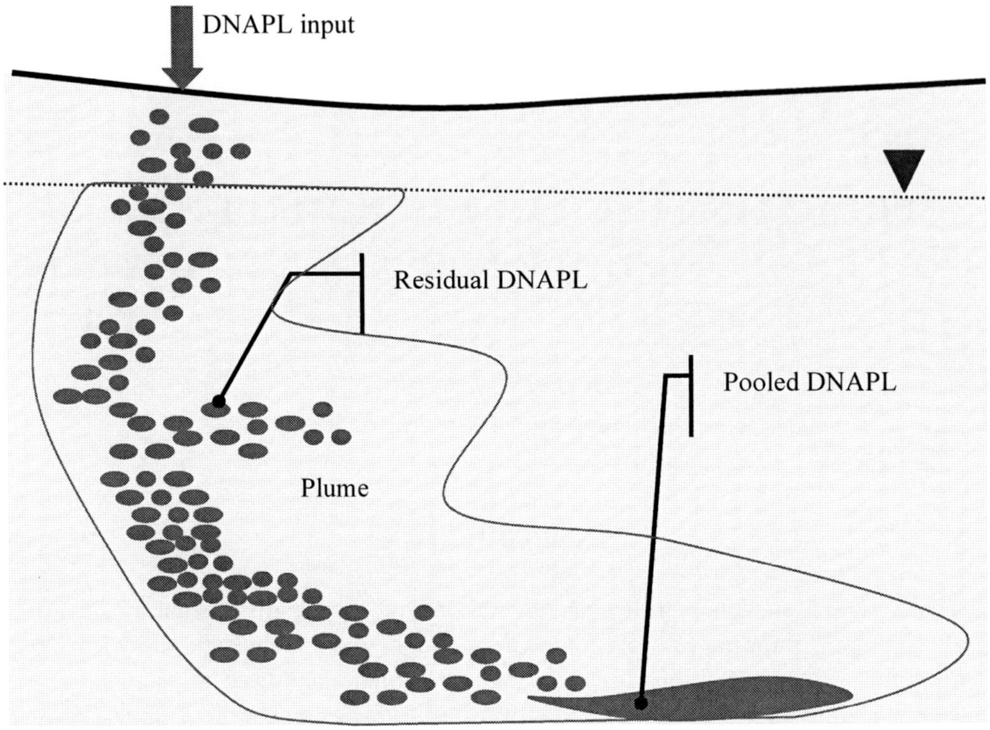

Figure 2. Cartoon showing the behaviour of DNAPL in the sub-surface environment.

Typical DNAPLs of environmental concern include chlorinated hydrocarbon, solvents such as chloroform and trichloroethene though this list can be extended to include a wide variety of halogenated organic liquids; coal tars and creosotes; and transformer and insulating oils containing polychlorinated biphenyls. The fundamental physical and chemical properties that exert a key influence on their behaviour in the sub-surface include: density; interfacial tension; wettability; viscosity; solubility; vapour pressure and chemical composition

Restoration of contaminated aquifers to drinking water standards demands: the use of effective site investigation tools to delineate the extent and location of contaminant zones; an understanding of the behaviour of organic liquids in aquifer formations and the use of remedial treatments that can either remove DNAPL pools and residuals or effectively contain/treat the contaminant plume.

DNAPL Behaviour in Sub-Surface Environments

As liquids, DNAPLs readily move both vertically and laterally in porous media such as unconsolidated deposits and fractured rock and because their specific gravities are greater than 1, they are able to enter the saturated zone where they will exist as separate liquid phases. Though immiscible with water, DNAPL components are sparingly soluble in water and these are slowly transferred from the DNAPL to flowing water to create an aqueous phase contaminant plume. The cartoon presented in figure 2 shows in a somewhat exaggerated form this behaviour.

DNAPL transport is dominated by gravitational and capillary forces. It has been noted by many researchers that droplets are detached from the trailing edge of a DNAPL body[1]. The formation of this residual DNAPL takes place as the result of pore-scale hydrodynamic instabilities.[2] These residual droplets are trapped in pore spaces by capillary forces as disconnected blobs and ganglia (as shown in figure 2). Normally the trapping capillary forces are much larger than the viscous forces generated by flowing groundwater, under natural hydraulic gradients and are thus immobile. Even the large hydraulic gradients generated by pumping are usually ineffective in mobilising residually trapped DNAPL. This therefore explains why groundwater remediation schemes based upon pumping are very often ineffective[1].

The transport of DNAPLs in sub-surface environments is complex. This complexity arises as the result of geological heterogeneity and the specific properties of the organic liquid. A key feature of DNAPL transport is the tortuous nature of the pathway it takes as it moves. This tortuosity arises because the entry of DNAPL into a specific pore depends upon a number of factors. These include the pressure exerted by the DNAPL (in figure 3 this is designated as P_{DNAPL}). To enter the pore depicted in the figure 3 the DNAPL must pass through the pore throat (this is the point where the grains come into contact with each other). Whether the DNAPL is able to pass into the pore depends upon the balance between P_{DNAPL} and P_W (water pressure). DNAPL entry into the pore necessarily entails the displacement of water. Water displacement is resisted by the water pressure. It is easily shown that pressure (P) exerted by a fluid is given as:

$$P = \rho g h \tag{1}$$

where ρ is the density of the fluid, g is the acceleration due to gravity and h is the height of the continuous fluid column (head). Clearly as the height of the DNAPL column is greater than that of water; and since the density of DNAPL is greater than that of water entry should be facile. The pressure difference $P_{DNAPL} - P_W$ is referred to as the capillary pressure. Pore entry is, however, also opposed by the capillary forces generated at the pore throat. Capillary forces are invariably produced at the curved interface between two immiscible fluids and they manifest themselves as a pressure jump across the interface. In figure 3 the pressure at A – immediately situated on the DNAPL side of the curved interface is greater than the pressure at B immediately situated on the water side of the interface (if the interface were flat they would be equal). The pressure at C (which is at the same elevation as B) is greater than the pressure at B and this pressure difference provides an additional pressure that tends to prevent DNAPL pore entry. The magnitude of the pressure difference (referred to as the pore entry pressure, P_e) is given by the Laplace equation[3]

$$P_e = \frac{2\sigma \cos\theta}{r} \qquad (2)$$

where σ is the interfacial tension between the DNAPL and water; θ is the contact angle and r is the pore throat radius.

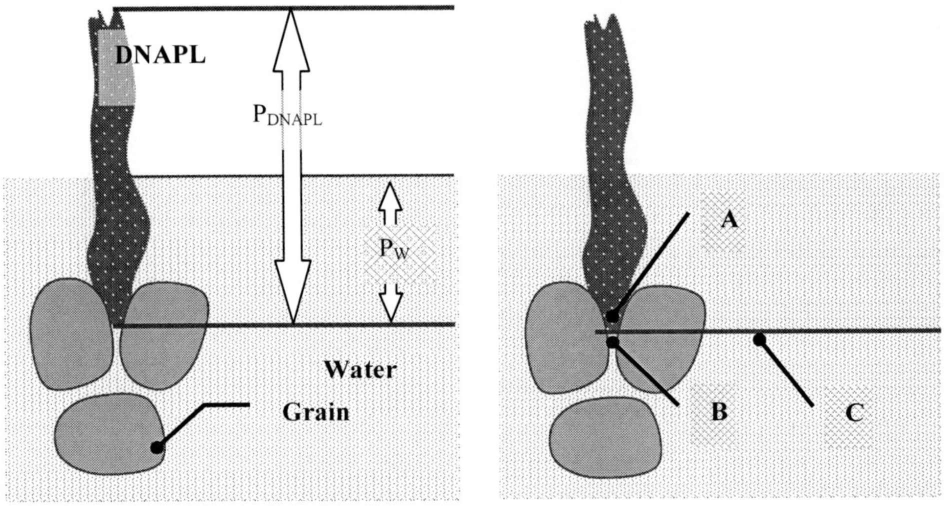

Figure 3.Cartoon showing the issues that affect the entry of DNAPL into a pore.

An interfacial tension invariably develops at the interface between two immiscible liquids. Figure 4 provides, in cartoon form, an explanation as to why, at the molecular level, water is immiscible with non-polar organic liquids. The transfer of a non-polar molecule from its own bulk phase to the aqueous phase can be thought of as involving the following conceptual stages[4]. Initially a vacancy or hole must be created in the water phase. Conceptually this involves the removal of a water molecule and the concomitant breaking of the intermolecular bonds with its near neighbours. The hole is then occupied by a non-polar molecule and this forms new intermolecular bonds with its near neighbours. Immiscibility arises because the water-water intermolecular bonds are stronger than intermolecular bonds between water molecules and the non-polar molecule. As a consequence, mixing, at the molecular level, carries with it a strong free energy penalty.

This free energy penalty also develops at the interface between two immiscible liquids such as water and a DNAPL for precisely the same reasons. At the molecular level, as shown in the cartoon, water-water bonds are necessarily broken to form the water DNAPL interface. This gives rise to a free energy penalty that results in a strong tendency for the water molecules to withdraw from the interface (shown in the cartoon in figure 5 as a thick arrow). This tendency in turn gives rise to a strong contractile force at the interface, which effectively seeks to minimise the size of the interface and is designated the interfacial tension.

The contact angle describes the ability of a liquid to spread across a solid surface in competition with another immiscible fluid (which can be air or another liquid). The diagram in figure 6 shows how the contact angle is defined. In this particular case the contact angle made by the DNAPL drop is less than 90° indicating that in the presence of water DNAPL does not spread (or wet) the solid surface. In strongly water wetting media such as aquifer deposits and rocks, DNAPLs invariably have contact angles that are quite low.

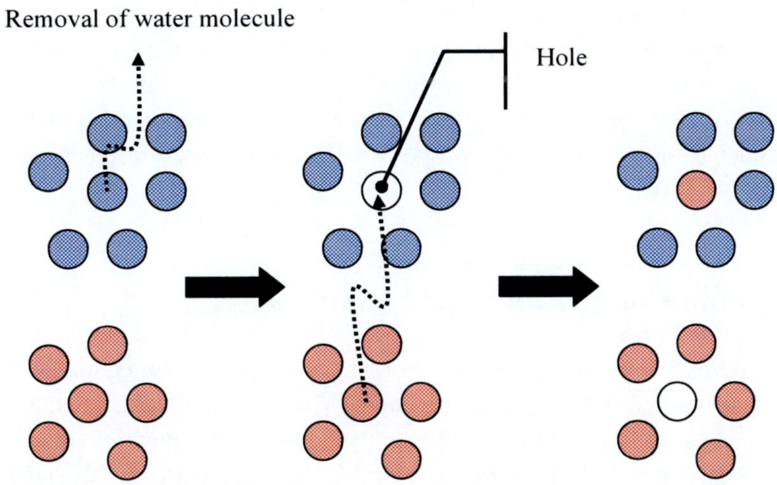

Figure 4. A conceptual drawing showing the stages by which a non-polar molecule (shown in red) is transferred from its own bulk phase to the aqueous phase (blue).

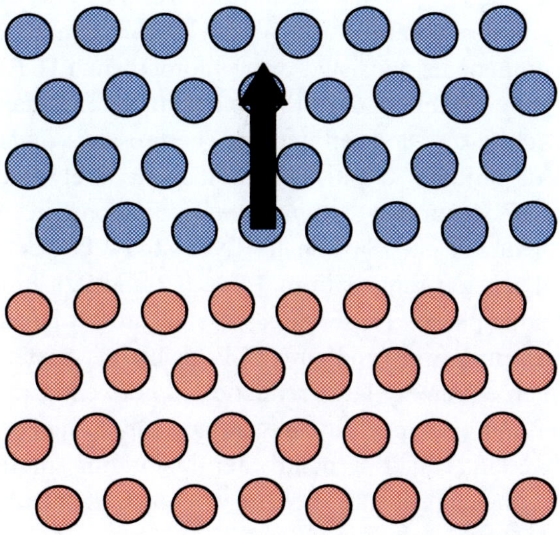

Figure 5.

Pore entry is thus facilitated by: a low interfacial tension; a large pore throat radius; and/or a contact angle close to 90° (at 90° the entry pressure is of course zero). It is the variations in entry pressure that provides the explanation for the tortuous pathways that DNAPLs take in the sub-surface. Moreover at the boundary between coarse grained and fine grained sub-surface horizons, DNAPL will move laterally and may accumulate at these boundaries as pools.

In fractured rock DNAPL pathways are dictated by the nuances of fracture connectivity and fracture aperture in the same way. It is commonly assumed that in dual porosity systems entry pressures for fractures are lower than the entry pressures for the primary porosity in the sedimentary deposits and that as a consequence DNAPL tends to move through interconnected fracture network. It is equally important to recognise that the forces responsible for the trapping of DNAPL residuals in pores and fractures are also capillary forces and that as a consequence the size of trapping forces are proportional to interfacial tension and contact angle and inversely proportional to either pore throat radius or fracture aperture.

DNAPLs are Multi-Component Liquid Mixtures

It is unlikely that DNAPLs found in sub-surface environments consist of a single component. Coal tar and creosote NAPLs are derived from coal and as such consist of a wide variety of polyaromatic hydrocarbons (PAHs), phenolic and heterocyclic compounds[5]. Chlorinated hydrocarbon solvents (CHSs) are widely used in many industrial and cleaning processes such as vapour degreasing of metal components in the engineering industry; as a solvent in the chemical and pharmaceutical industry or in dry cleaning. As a consequence used CHSs are likely to contain a wide number of components. Moreover co-disposal of CHS wastes with other effluent streams from on-site processes is likely to create mixtures with extremely complex and heterogeneous compositions. Powers and co-workers[6] have characterised DNAPL samples obtained from the US Department of Energy's Savannah River Site. The major components consisted of trichloroethene (TCE) and tetrachloroethene (PCE). However they were also able to detect trace quantities of metals; high molecular mass hydrocarbons and fatty acid esters and fatty acids. The presence of fatty acids is potentially significant since in their de-protonated/ionised form they are likely to exhibit surface activity. The major manifestations of surface activity are: reductions in interfacial tension and alterations in wetting. Thus in the Savannah River study Powers *et al*[6] reported that water/DNAPL interfacial tension ranged from 1.2 ± 0.2 and 2.3 ± 0.3 mN m^{-1}, depending upon the pH of the aqueous phase. These values are an order of magnitude lower than the interfacial tension between pure water and pure TCE – which is quoted as 44.2 ± 0.4 mN M^{-1}. Such reductions indicate an extremely large reduction in pore entry pressure and suggest that DNAPL is likely to enter and be trapped in fine grained sub-surface layers which will render them difficult to remove and make groundwater restoration more problematic. In our laboratories, we too have measured water/field recovered CHS DNAPL interfacial tensions and have obtained values that range between 8 and 10 mN m^{-1}.

Figure 7 is a diagrammatic representation and description of what is described as surface activity. Molecules, such as fatty acids (hexadecanoic acid is shown in the diagram) exhibit surface activity, as a result of their specific molecular structure. The carboxylate end of the molecule is charged and is thus hydrophilic the remainder of the molecule comprises an alkyl chain and is therefore hydrophobic. Such molecules are readily adsorbed at the interface between water and non-polar organic liquids such as DNAPLs. The head group (shown in yellow in figure 7) is readily absorbed by water. However the alkyl chain (the black lines in figure 7) is repelled by water and is more readily absorbed by the DNAPL. The phase preferences of both parts of the fatty acid are thus readily met by adsorption at the interface.

This preferential adsorption at the interface permits water molecules to return to the bulk and as a consequence the interfacial tension which is – as maybe recalled – a measure of the free energy penalty associated with the formation of the interface, is reduced.

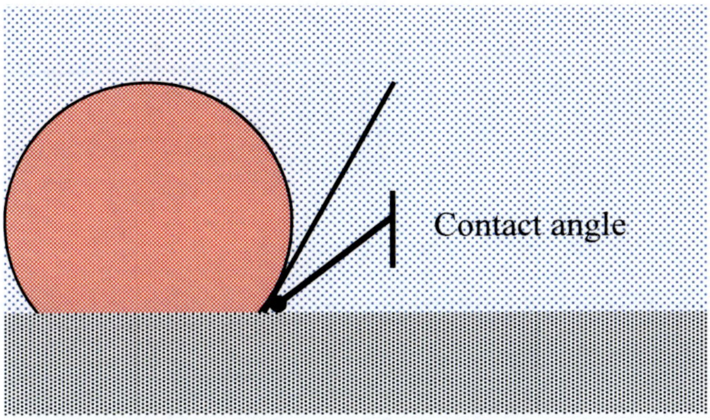

Figure 6.

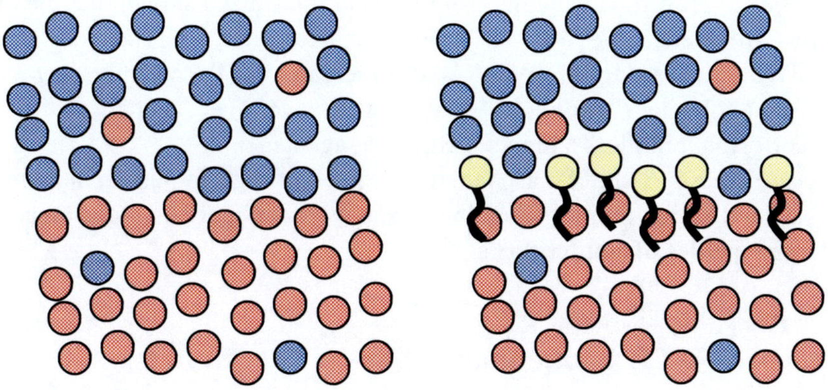

Figure 7. Cartoon showing the adsorption of surface active compounds (yellow) at the interface between water (blue) and DNAPL (red).

Figure 8 (based upon the work of Zheng and Powers[7]) provides, in cartoon form, an example of how the wetting preference of the mineral surface can be altered by the presence of charged surface active compounds in the NAPL. As the NAPL droplet approaches the surface the film of water that separates the NAPL and surface becomes thinner. For the wetting preference to alter, this film needs to rupture, which as a result permits the NAPL to directly contact the surface. This will happen in the example shown in figure 8 because of the attractive electrostatic forces operating between the negatively charges solid surface and the positively charged head groups of the surface active solutes in the DNAPL. As the wetting fluid NAPL will spread over the surface as a film separating the water and mineral surface. This change in NAPL architecture will modify the dynamics of NAPL component dissolution in groundwater[8] by effectively increasing the interfacial contact area and thereby increasing the flux of NAPL components from the NAPL phase to the aqueous phase.

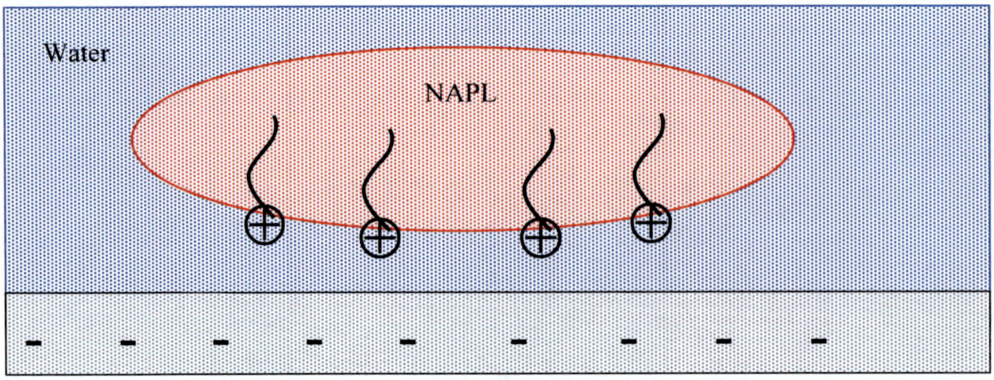

Figure 8. The wetting preference of the mineral surface can be altered by the thinning and ultimate rupture of the water film separating the NAPL droplet from the surface. Attractive electrostatic forces as shown help this process.

Even before use, CHSs employed in industry contain additives. These additives provide solvent formulations that are stabilised and specifically adapted to the challenges of the industrial processes they are being used for. Thus TCE maybe formulated with trace amounts of trimethylamine, to prevent solvent degradation[9] and it has been shown that the presence of trimethylamine will reduce the water/TCE interfacial tension[10]. Surfactants are added to chlorinated solvents used in dry cleaning[11] and these can alter the wetting characteristics of rocks from strongly water wetting to weakly water wetting[12]. Other additives include antioxidants, corrosion inhibitors and metal chelating agents.

Coal tars and creosotes contain asphaltenes. These are high molecular mass compounds that possess acid and basic groups and again these may affect interfacial tension against water and the wetting properties of aquifer rocks[7,13,14]

Remediation of DNAPL Contaminated Sub-Surface Systems

It is important that the reader appreciates that technical solutions that focus on DNAPL removal are procedures that are solely concerned with the displacement of the pooled and residually trapped separate organic liquid phase. On the other hand technologies that treat the plume are effectively based upon either physical techniques that control or manage groundwater flow or chemical techniques that degrade or otherwise remove the contaminants dissolved in the flowing groundwater. Thus remediation of DNAPL impacted sub-surface environments can be conveniently divided into technologies that seek to remove the separate organic liquid phase and remedial systems that control, manage and/or degrade the contaminants encountered in the plume. It must be emphasised that technical solutions that are designed to treat the plume are neither necessarily nor usually effective against DNAPL source zones. The rest of the chapter will examine bulk displacement of DNAPL.

POOLED DNAPL REMOVAL

The term pooled DNAPL (sometimes referred to as free phase) describes a large accumulation of interconnected DNAPL which is either mobile or potentially mobile. Pools of DNAPL accumulate at the interface between a relatively permeable layer and a lower less permeable layer. In fractured rock pooled DNAPL tends to be encountered in horizontal or sub-horizontal fractures[2]. If site investigation reveals the presence of mobile DNAPL then it is important that the DNAPL body is stabilised (i.e. the DNAPL body is immobilised). It is possible to remove pooled DNAPL by water flooding. Effectively the hydraulic gradient is locally raised via the use of water injection wells hydraulically linked to water extraction wells[2,15] as shown in figure 9.

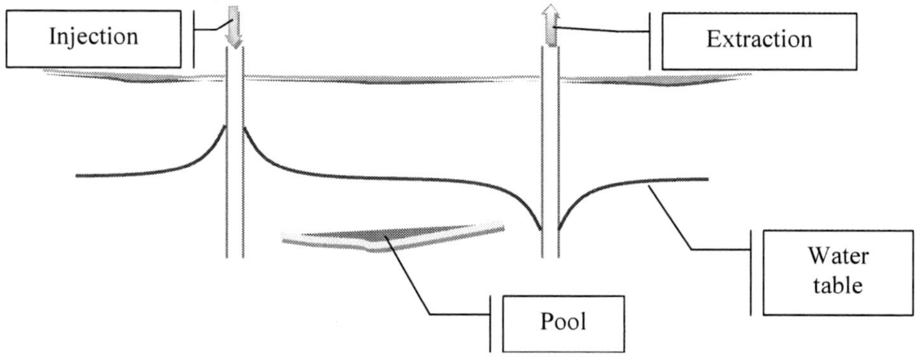

Figure 9. Schematic diagram of water flooding as a remedial tool to remove pooled DNAPL.

For viscous DNAPLs such as coal tar and creosote, water flooding may not be effective due to the development of viscous instabilities at the displacement front. These instabilities arise because water as the displacing liquid has a lower viscosity than the coal tar/creosote DNAPL. As a result of these instabilities, water may completely bypass the DNAPL pool[15]. This problem may be obviated by the use of a viscosifier – in the aqueous phase – such as xanthan gum that will increase the viscosity of the displacing aqueous phase. Giese and Powers have shown that this has the desired result in NAPL wetting systems but seems to make little difference in water wetting systems[15].

REMOVAL OF RESIDUALLY TRAPPED DNAPL

DNAPLs such as coal tar that have densities only slighter greater than water but viscosities much greater than that of water, may be mobile for decades in sub-surface environments[2]. However CHS DNAPLs that have high densities and low viscosities are likely to be encountered as trapped residuals. As we have already seen these trapping forces are capillary forces. The critical conditions for mobilisation are dependent upon a combination of the viscous and buoyancy forces that act against the trapping capillary forces. Environmental engineers identify two dimensionless numbers that can be used to define the critical conditions necessary to displace the trapped DNAPL. The Bond Number (N_B) relates buoyancy forces to capillary forces

$$N_B = \frac{\Delta\rho g k_i k_{rw}}{\sigma \cos\theta} \tag{3}$$

$\Delta\rho$ is the density difference between the DNAPL and water, g is the acceleration due to gravity, k_i is the intrinsic permeability of the porous medium (this is related to the distribution of pore sizes in the porous medium, k_{rw} is the relative permeability of water (this describes the ability of water to move through the porous medium given that some potential pathways are not permitted since these contain trapped DNAPL), σ is the interfacial tension between the DNAPL and water and θ is the contact angle.

The capillary number (N_{Ca}) relates viscous forces to capillary forces and is defined as:

$$N_{Ca} = \frac{q_w \mu_w}{\sigma \cos\theta} \tag{4}$$

q_w is the Darcy velocity of the aqueous phase (this is the velocity of water through unit cross sectional area of porous medium under a specific hydraulic gradient) and μ_w is the viscosity of the aqueous phase. Both numbers can be combined in a trapping number index, (N_T) which for horizontal flow is given as[16]:

$$N_T = \sqrt{N_B^2 + N_{Ca}^2} \tag{5}$$

and for vertical flow the expression becomes[16]:

$$N_T = \left| N_B + N_{Ca} \right| \tag{6}$$

DNAPL is displaced when the value of N_T is greater than some critical value. Inspection of the equations 4 and 5 show that the values of both N_B and N_{Ca} can be increased by reducing the interfacial tension or increasing the contact angle. In addition N_{Ca} can be increased by increasing either the viscosity of the aqueous phase or the Darcy velocity.

The interfacial tension is readily lowered through the use of surfactants. However it is important to appreciate that the reductions necessary for mobilisation are not necessarily achieved by using a simple single component aqueous surfactant solution. It is a well documented feature of surfactants that as the bulk aqueous concentration of surfactant is increased then either the interfacial or surface tension is reduced. However a lower limit is reached wherein additional surfactant above a specific concentration begins to undergo self assembly to form micelles (see figure 10).

Micelles are surfactant aggregates in which the hydrophobic non-polar chains self assemble in order to reduce their exposure to water, whilst the polar head groups form a palisade layer fully exposed to water. The resultant structure is a colloid of nano-sized dimensions which is readily dispersed in the aqueous phase. Micelles behave as a separate phase (they are very often referred to as a pseudo phase) in that the presence of micelles dispersed in the aqueous phase results in an aqueous phase that is saturated with unimeric surfactant (see figure 10). The concentration of unimeric surfactant in equilibrium with the micellar pseudo phase is identified as the critical micelle concentration (cmc).

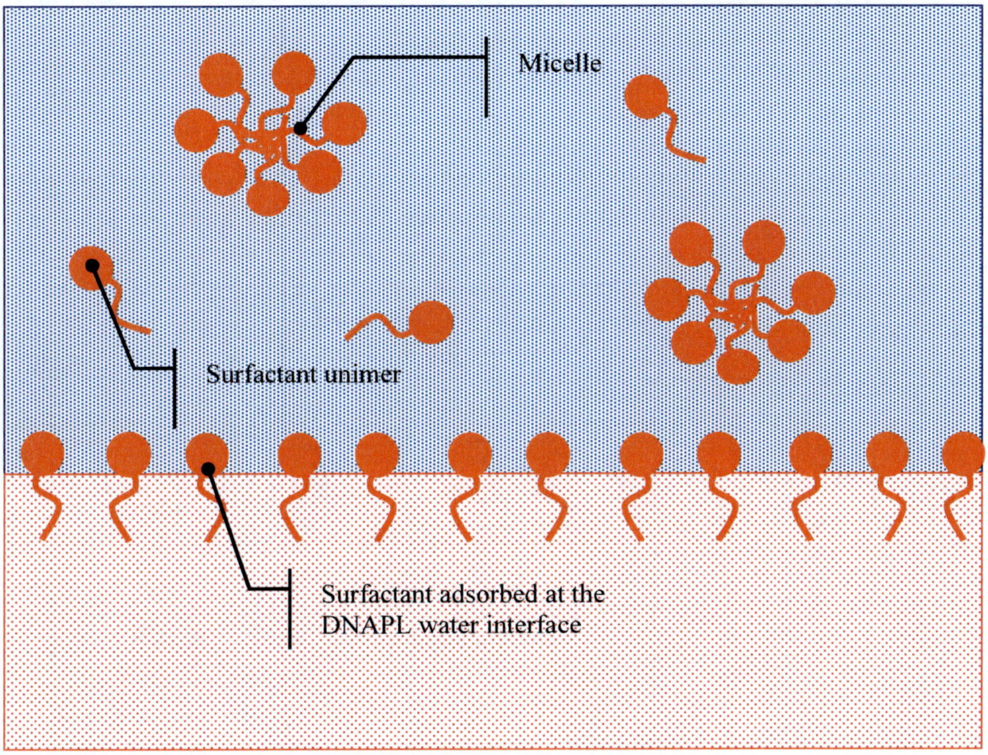

Figure 10. Surfactant distribution in a DNAPL water system.

The structure of a micelle can be viewed as comprising an oil phase located in the core surrounded by a shell of polar/ionic groups that favourably interact with water. The shell ensures that micelles are readily dispersed in the aqueous phase. Hydrophobic solutes are readily incorporated into the micelle core which has the effect of increasing their apparent aqueous solubility. The process is referred to as solubilisation.

Experience acquired in the petroleum industry suggests that the displacement of an organic liquid, by water, in a porous medium is achieved by reducing the interfacial tension to ultra-low levels[17]. This is realised by carefully adjusting the composition of the displacing surfactant solution. This can be accomplished through the use of surfactant mixtures. One example is provided by the work of Pennell and co-workers[18] who examined the surfactant enhanced remediation of TCE emplaced in sand columns. They found that surfactant solutions consisting of a single surfactant (Tween 20 – concentration 4%) had an IFT of 5 mN m^{-1} against TCE; whereas a 4% solution consisting of a 1:1 aqueous mixture of sodium diamyl-sodium dioctyl sulphosuccinate had an IFT of 0.09 mN m^{-1} against TCE.

An alternative route to producing systems with ultra low interfacial tensions is to produce micro-emulsions. Micro-emulsions unlike macro-emulsions are thermodynamically stable and are spontaneously formed under specific compositional conditions. The phase preference of a surfactant is determined by the balance between the hydrophilic and lipophilic forces in the surfactant molecule. The phase preference is indicated by the HLB (hydrophile lipophile balance) number. Surfactants with high HLB numbers are water soluble where as those with low numbers are oil soluble. The HLB number of a surfactant can be reduced through the addition of salt to the aqueous phase. This has the effect of driving increasing amounts of surfactant to the interface[19] which inevitably leads to increasing reductions in the interfacial tension (see figure 11). As the salinity of the aqueous phase is increased a point is reached where the surfactant systems shows an equal preference for both the aqueous phase and oil phase[19]. At this point the surfactant is overwhelmingly located at the interface and since not all the surfactant can be accommodated at the interface we see the spontaneous formation of a new phase called a middle phase microemulsion or Winsor type 3 microemulsion (see figure 11). It is entirely possible to adjust the compositional features of the system so that the oil and water phases are completely incorporated into the microemulsion.

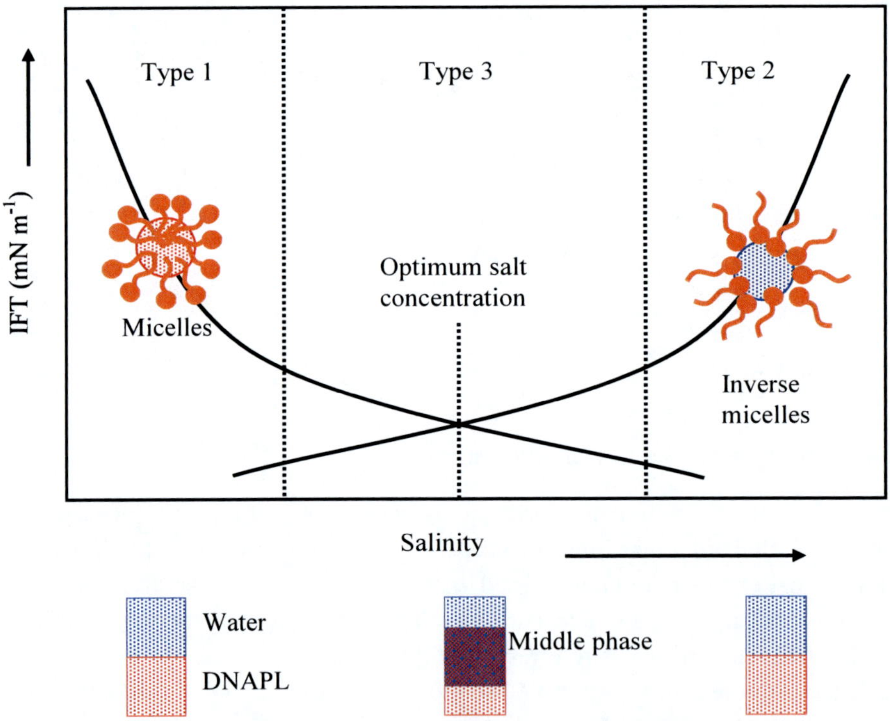

Figure 11. Winsor phase diagram based upon Rosen[17].

Figure 12 demonstrates that a middle phase formation is easily formed through the addition of salt to the aqueous phase. The aqueous phase consists of sodium dodecyl sulphate (SDS) and varying quantities of sodium chloride. At low salt concentrations the aqueous phase consists of micelles which solubilise the coal tar. In the bottle on the left the aqueous phase is dark brown as a result of coal tar solubilisation in SDS micelles. In the middle bottles a middle phase microemulsion is formed. The colour of the aqueous phase has altered as a result of the transfer of surfactant and coal tar from the aqueous phase to the middle phase. Finally in the right hand bottle the salt concentration is so high that SDS is now soluble in the coal tar phase. The middle phase in this system has disappeared.

There are, however, problems associated with middle phase formation associated with the risk of gravity driven mobilisation of the DNAPL to deeper previously uncontaminated parts of the aquifer. It is possible to obviate this problem through the use of an alcohol that preferentially partitions from the aqueous displacing solution to the DNAPL thereby making the NAPL phase neutrally bouyant[20]. Sabatini *et al.*[19] have shown that an alternative approach adjusts the composition of the aqueous surfactant phase with sufficient salt such that considerable amounts of TCE DNAPL can be solubilised (the authors describe the process as super-solubilisation) but the interfacial tension of the system is high enough so that the capillary number for the system is below the critical number for mobilisation.

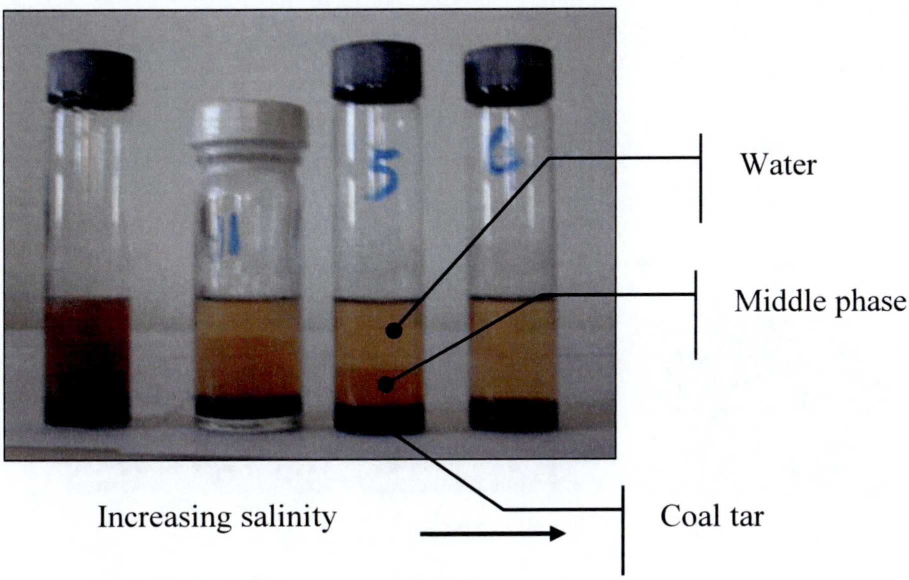

Figure 12. Middle phase formation in a coal tar water system.

DNAPL REMOVAL AND WETTING PROPERTY CHANGES

It is possible that for complex DNAPL mixtures, which contain surface active components, the DNAPL may have altered the wetting properties of the sub-surface sediment or rock surfaces, with which it has come into contact with, from strongly water wetting to oil wetting. Experience gathered by researchers in the petroleum industry suggests that under these conditions oil removal is favoured by a change in wetting especially if the oil is retained in oil wetting low permeability layers[21]. In this case a change from oil wetting to water wetting conditions permits the spontaneous imbibition of water into the oil occupied pores and the concomitant displacement of the oil. Surfactants can be used to alter the wetting properties of a solid surface however little work has been reported on wettability manipulation as mechanism for the removal of trapped DNAPL. One investigation undertaken in our laboratories showed that surfactants can be used to aid coal tar removal by altering the wetting preferences of quartz[22].

In figure 13 a series of photographs show how a drops of coal tar that have been placed upon a quartz surface behave when a portion of the drop is withdrawn using a syringe. In photographs 13a – c the concentration of surfactant present in the surrounding aqueous phase is 5×10^{-5} %. It is evident that under these conditions the coal tar drop is firmly attached to the quartz surface. This is clearly shown in figure 9b where it is apparent that the drop resists withdrawal to such an extent that a considerable part of the drop becomes detached and remains on the quartz surface (figure 13c). For photographs 13d – f the surfactant concentration was 5×10^{-3} %. In this case the resistance to withdrawal is less and only a small portion of the drop remains attached to the surface (figure 13f). For the final series of photographs 13g – i the surfactant concentration is 1×10^{-2} %. The drop in this system is readily detached from the surface. The photographic data provides clear evidence of a change in the adhesional wetting of the coal tar to the quartz surface. The work of adhesion (W_a) is defined as the work necessary to separate unit are of liquid from the solid surface and is given by the following expression:

$$W_a = \sigma(1-\cos\theta) \tag{7}$$

In this system the surfactant reduced the work of adhesion by reducing the surface tension and by increasing the contact angle see figure 14.

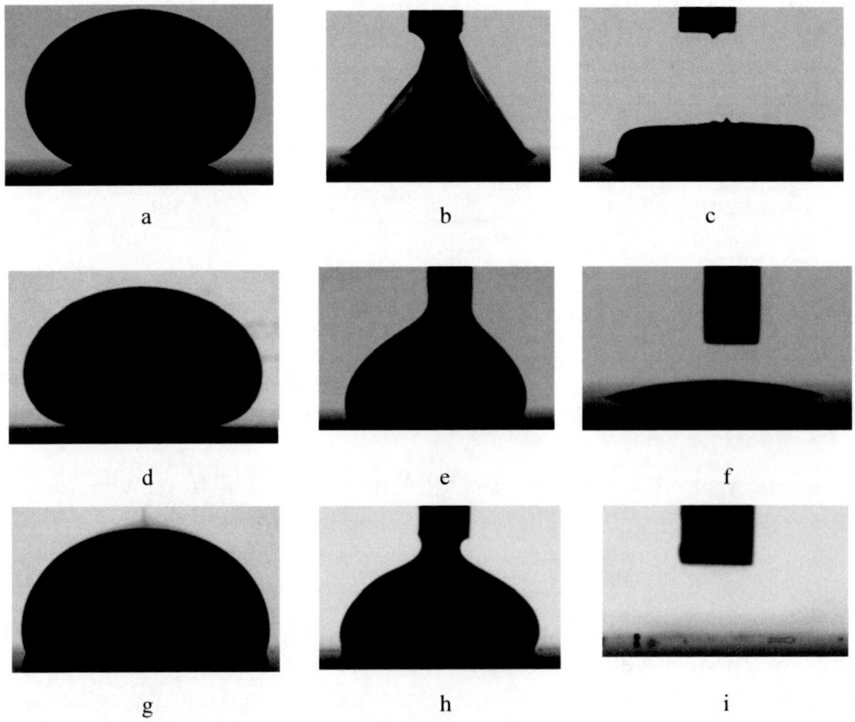

Figure 13. A series of photographs show how coal tar can alter the wetting properties of a quartz surface and how the presence of surfactant can alter these wetting changes.

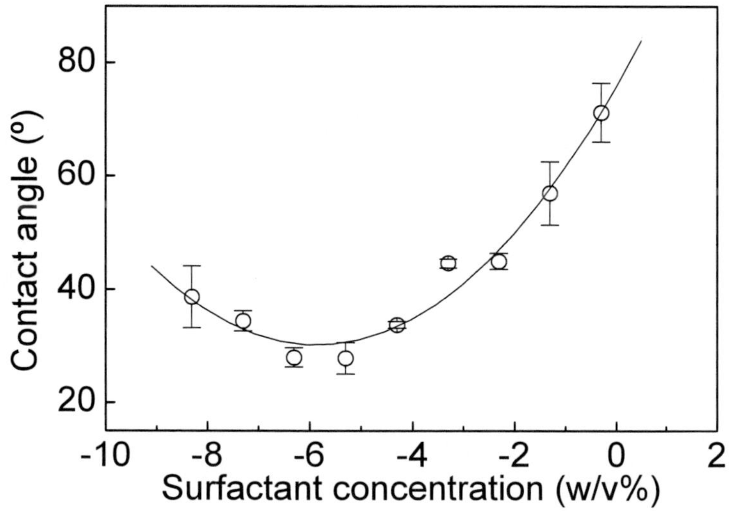

Figure 14. Surfactants can be used to readily alter the contact angle.

REMEDIATION BASED UPON SURFACTANT SOLUBILISATION

Many workers suggest that a more risk adverse remedial technology relies upon solubilisation in surfactant micelles. The graph in figure 15[23] shows the increase in the apparent aqueous solubility of naphthalene as a function of surfactant (Pluronic P103) concentration. Solubility is shown as solubility enhancement – the ratio of the apparent aqueous solubility in the surfactant solution to the solubility in water. It is evident that the presence of surfactant increases apparent solubility even below the cmc where only surfactant unimers are present. As shown in figure 16, this is presumably due to surfactant-solute interactions. Above the cmc the rate at which the solubility enhancement increases with surfactant concentration is improved as a result of solubilisation Surfactant sub-surface flushing is performed using injection wells and extraction wells using a configuration like that shown in figure 9. One particular problem that may emerge is the loss of surfactant by adsorption. Adsorptive losses may arise because of surfactant head group attachment to soil or rock surfaces. This is normally avoided by using anionic surfactants such as SDS which, because of its negative head group, should be repelled by negatively charged surfaces. However small adsorptive losses have been noted, by workers,[24,25] of anionic surfactants to soils and sediments. In figure 15 it is suggested that surfactants may also be lost *via* sorption to soil organic matter.

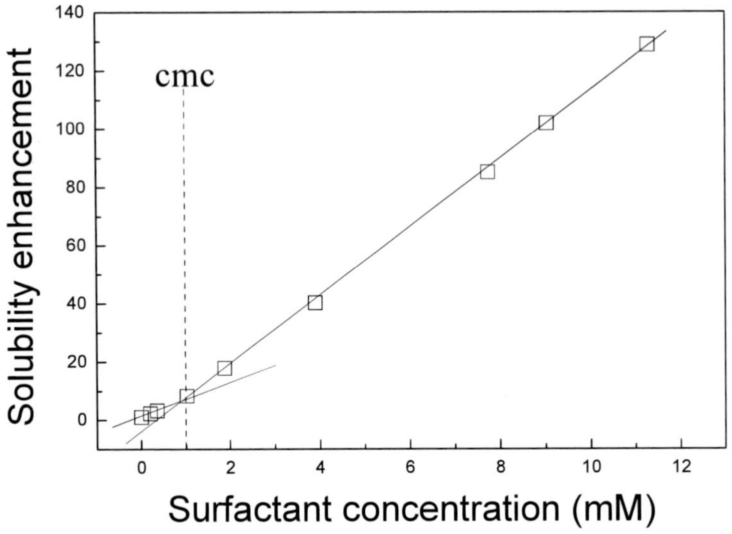

Figure 15.

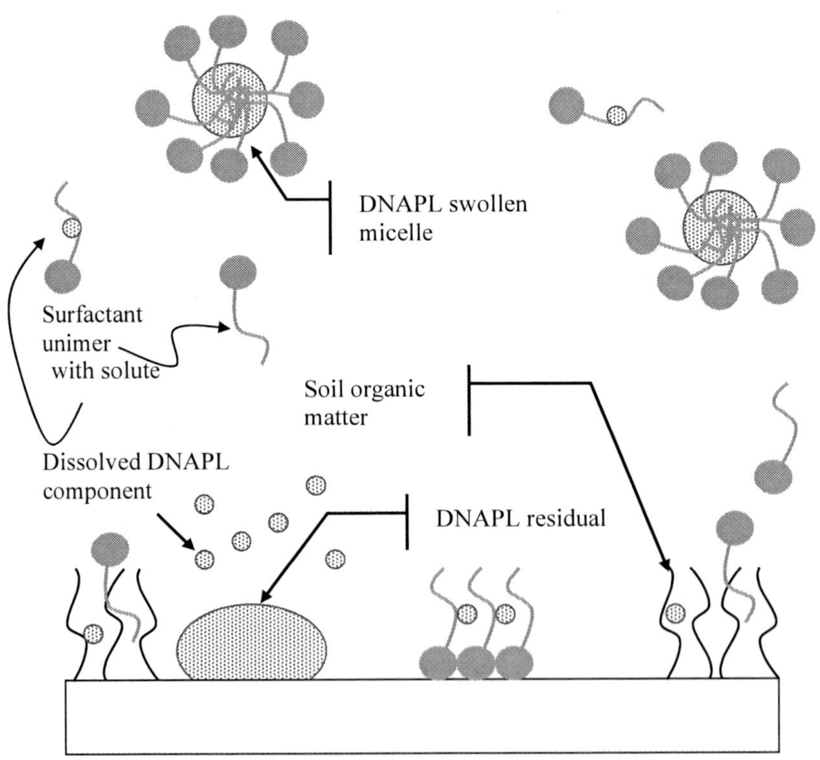

Figure 16. Diagrammatic representation of surfactant solubilisation.

Very often for non-ionic surfactants the hydrophilic part of the molecule is composed of a variable number of ethylene oxide units (shown in the figure). The presence of these groups provides a mechanism by which these surfactant molecules can be adsorbed, *via* hydrogen bonding interactions, to surface silanol groups on silica phases.[26]

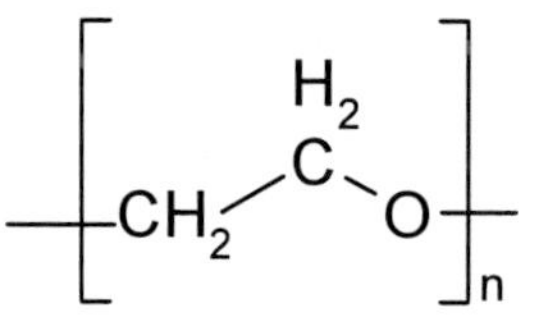

SURFACTANT RECYCLING AND RE-USE

Surfactants are expensive and as a consequence it is necessary to devise some way of capturing them from the waste stream and reusing them[27]. Two stages are necessary to achieve this goal: (i) separation of surfactant and solubilised component; and (ii) separation of surfactant from the waste stream obtained from the extraction wells. Separation of micelles from the aqueous waste stream is readily accomplished by micellar enhanced ultrafiltration[28]. For volatile solubilised components, like CHSs or aromatic hydrocarbons it is possible to use vapour stripping to separate solute and surfactant[29]. There are, however, two technical problems associated with vapour stripping[29]. These are: (i) foaming - the presence of surfactants in the aqueous phase support the formation of foams when air is passed through and (ii) the presence of surfactants hinder the transfer of volatile organic compounds from the micellar pseudo-phase to the vapour phase. The use of hollow fibres can mitigate these problems[29].

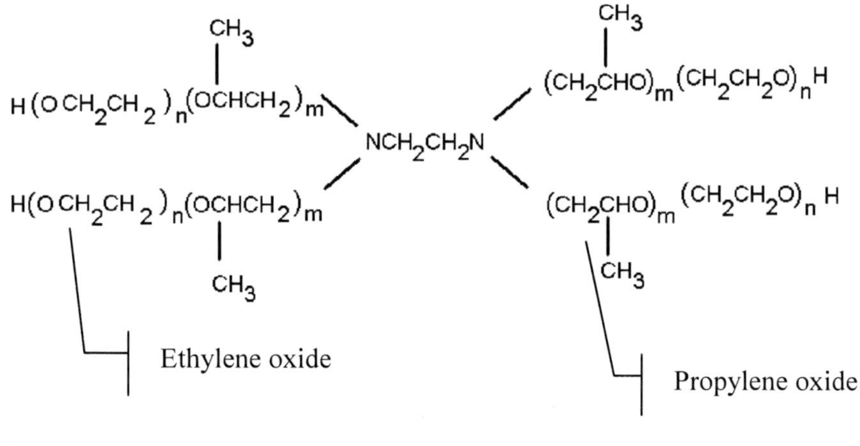

Figure 17. The molecular structure of the poloxamine block copolymers.

In our laboratories we have investigated the use of poloxamine block copolymeric surfactants as potentially useful solubilisation agents.[30] The general molecular structure of the poloxamines is shown in figure 17. The amphiphilic nature of these molecules is derived from the combination of the hydrophilic ethylene oxide blocks and the hydrophobic propylene oxide blocks. The balance between the hydrophobic properties of the molecule and its hydrophilic properties is controlled by altering the size of the blocks. Since these molecules are large they should be readily removed from aqueous waste streams by membrane filters.[31] At room temperature these molecules are water soluble. However these molecules show the interesting property of forming micellar like aggregates at higher temperatures. This occurs because of the development of increasing hydrophobicity in the propylene oxide block as the temperature rises[32]. These molecules also aggregate as the concentration of copolymer is increased.[33] These aggregates very readily solubilise hydrophobic solutes.[33] Interestingly poloxamines possess a central ethylene diamine group which is able to bind two protons[34] to form a positively charged molecular species. Under acid conditions protonation occurs in the middle of the propylene oxide block. This has the effect of preventing aggregation[35] and thus affecting the solubilisation capacity of the system.[33] Poloxamines are used in a variety of pharmaceutical applications[36] and they are therefore unlikely to be environmentally dangerous. They should be readily trapped by ultrafiltration units[37] and the solubilisation can be readily reversed by a reduction in pH.[33] It is therefore proposed that these molecules could be extremely suitable for use in sub-surface surfactant enhanced remediation.

Concluding Remarks

The object of this chapter has been to provide an overview of how surfactants can be used for the restoration of DNAPL contaminated sub-surface environments. The reader is cautioned that successful implementation of this technology demands a good understanding of the contaminated site; an adequate characterisation of site sub-surface geology and hydrogeology; a good indication of where the DNAPL source zone is and how much DNAPL is present.

I would like to thank Dr. Luis Torres (IPN) Dr. Erick Bandala (UDLA) and Dra. Rosario Iturbe (UNAM) for their friendship and kindness; and the Royal Society and Academia Mexicana de Ciencias for funding.

References

[1] Mackay, D. M.; Cherry, J. A. *Environ. Sci. Technol.* 1989, *23*, 630.

[2] Kueper, B. H.; Wealthall, G. P.; Smith, J. W. N.; Leharne, S. A.; Lerner, D. N. *An illustrated handbook of DNAPL transport and fate in the subsurface (For environmental regulators, consultants, and engineers)*; Environment Agency, 2003.

[3] Hunter, R. J. *Foundations of Colloid Science*; Oxford University Press: Oxford, 1995; Vol. 1.

[4] Schwarzenbach, R. E.; Gschwend, P. M.; Imboden, D. M. *Environmental Organic Chemistry*; John Wiley & Sons: New York, 1993.

[5] Melber, C.; Kielhorn, J.; Manglesdorf, I. *Concise International Chemical Assessment Document 62: Coal Tar Creosote*; World Health Organisation: Geneva, 2004.

[6] Dou, W.; Omran, K.; Grimberg, S. J.; Denham, M.; Powers, S. E. *J. Contam. Hydrol.* 2008, *97*, 75.
[7] Zheng, J.; Powers, S. E. *J. Contam. Hydrol.* 1999, *39*, 161.
[8] Zhao, W.; Ioannidis, M. A. *Adv. Water Resour.* 2007, *30*, 171.
[9] Jackson, R. E.; Dwarakanath, V. *Ground Water Monit. Remediat.* 1999, *19*, 102.
[10] Harrold, G.; Gooddy, D. C.; Reid, S.; Lerner, D. N.; Leharne, S. A. *Environ. Sci. Technol.* 2003, *37*, 1919.
[11] Hoggan, J. L.; Bae, K.; Kibbey, T. C. G. *J. Contam. Hydrol.* 2007, *93*, 149.
[12] Harrold, G.; Lerner, D. N.; Leharne, S. A. *J. Contam. Hydrol.* 2005, *80*, 1.
[13] Zheng, J.; Shao, J.; Powers, S. E. *J. Colloid Interf. Sci* 2001, *244*, 365.
[14] Zheng, J. Z.; Behrens, S. H.; Borkovec, M.; Powers, S. E. *Environ. Sci. Technol.* 2001, *35*, 2207.
[15] Giese, S. W.; Powers, S. E. *J. Contam. Hydrol.* 2002, *58*, 147.
[16] Pennell, K. D.; Pope, G. A.; Abriola, L. M. *Environ. Sci. Technol.* 1996, *30*, 1328.
[17] Rosen, M. J. *Surfactants and Interfacial Phenomena*, 2nd ed.; John Wiley & Sons: New York, 1989.
[18] Pennell, K. D.; Jin, M.; Abriola, L. M.; Pope, G. A. *J. Contam. Hydrol.* 1994, *16*, 35.
[19] Sabatini, D. A.; Knox, R. C.; Harwell, J. H.; Wu, B. *J. Contam. Hydrol.* 2000, *45*, 99.
[20] Lunn, S. R. D.; Kueper, B. H. *Environ. Sci. Technol.* 1999, *33*, 1703.
[21] Standnes, D. C.; Austad, T. *Colloid. Surface. A* 2003, *216*, 243.
[22] Dong, J. F.; Chowdhry, B.; Leharne, S. *Environ. Sci. Technol.* 2004, *38*, 594.
[23] Paterson, I. F.; Chowdhry, B. Z.; Leharne, S. A. *Langmuir* 1999, *15*, 6187.
[24] Westall, J. C.; Chen, H.; Zhang, W.; Brownawell, B. J. *Environ. Sci. Technol.* 1999, *33*, 3110.
[25] Ou, Z.; Yediler, A.; He, Y.; Jia, L.; Kettrup, A.; Sun, T. *Chemosphere* 1996, *32*, 827.
[26] Paterson, I. F.; Chowdhry, B. Z.; Carey, P. J.; Leharne, S. A. *J. Contam. Hydrol.* 1999, *40*, 37.
[27] Krebsyuill, B.; Harwell, J. H.; Sabatini, D. A.; Knox, R. C. Economic-Considerations in Surfactant-Enhanced Pump-and-Treat Remediation. In *Surfactant-Enhanced Subsurface Remediation*; ACS: Washington DC, 1995; ACS Symposium Series Vol. 594; 265.
[28] Lipe, K. M.; Sabatini, D. A.; Hasegawa, M. A.; Harwell, J. H. *Ground Water Monit. Remediat.* 1996, *16*, 85.
[29] Sabatini, D. A.; Harwell, J. H.; Hasegawa, M.; Robert, K. *Journal of Membrane Science* 1998, *151*, 87.
[30] Leharne, S.; Dong, J. Investigations of Surfactant Facilitated Removal of Coal Tar Contaminants From Manufactured Gas Works Soils. In *Groundwater Quality: Natural and Enhanced Restoration of Groundwater Pollution*; Thornton, S. F., Oswald, S. E., Eds.; IAHS, 2002 Vol. 275; 311.
[31] Hurter, P. N.; Hatton, T. A. *Langmuir* 1992, *8*, 1291.
[32] Armstrong, J.; Chowdhry, B.; Obrien, R.; Beezer, A.; Mitchell, J.; Leharne, S. *J. Phys. Chem.* 1995, *99*, 4590.
[33] Dong, J.; Chowdhry, B. Z.; Leharne, S. A. *Colloid. Surface. A* 2004, *246*, 91.
[34] Dong, J.; Armstrong, J.; Chowdhry, B. Z.; Leharne, S. A. *Thermochimica Acta* 2004, *417*, 201.

[35] Armstrong, J. K.; Chowdhry, B. Z.; Snowden, M. J.; Dong, J.; Leharne, S. A. *International Journal of Pharmaceutics* 2001, *229*, 57.

[36] Moghimi, S. M.; Hunter, A. C. *Trends in Biotechnology* 2000, *18*, 412.

[37] Choi, Y.-k.; Lee, S.-B.; Lee, D.-J.; Ishigami, Y.; Kajiuchi, T. *Journal of Membrane Science* 1998, *148*, 185.

In: Remediation of Soils and Aquifers
Editors: Luis G. Torres and Erick R. Bandala

ISBN 978-1-60741-372-1

Chapter 11

ACTIVE FAULTS AND AQUIFER VULNERABILITY: A CASE STUDY FROM CELAYA, MEXICO

J. A.Ramos-Leal, J.J. Carrera-Hernandez
Instituto Potosino de Investigacion Cientifica y Tecnologica

ABSTRACT

Three different methodologies to estimate aquifer vulnerability (SINTACS, AVI and Rn-222) were applied to the aquifer of Celaya, Mexico, located on a tectonically active zone in order to determine which of these methodologies better represents the existence of faults and fractures that increase the aquifer's vulnerability. The region where this aquifer is located contains structural alignments which need to be considered in a vulnerability study, as a large number of settlements in central Mexico are found on tectonically active zones (i.e. Morelia, Mexico City, Salamanca, Celaya, Irapuato, Aguascalientes and Queretaro). The comparison undertaken in this work, shows that the vulnerability indexes obtained with the aforementioned methodologies exhibit the impact of structural alignments differently: The SINTACS method does not show major vulnerability where these structures are found, whereas the AVI increases slightly in some of the regions where the alignment are located. The spatial distribution of the aquifer's vulnerability is better represented by the Rn-222 method, as the effects of the existing faults are reflected on the vulnerability maps developed with this methodology.

INTRODUCTION

The vulnerability intrinsic is a property of an aquifer system and depends on the susceptibility of the system to be adversely affected by any type of contaminant load, either from natural or anthropogenic origin (Vrba and Zaporec, 1994). This susceptibility depends on the inaccessibility of the saturated zone to surface contamination and on the attenuation capacity of the overlying strata of the aquifer system. However, in tectonically active zones, faults and fractures play an important role as they can breach through this protective system, thus increasing the risk of aquifer contamination.

These geologic structures are found in a large number of regions throughout the world and Mexico is not an exception, in particular on its central region (Charles and Kreitler, 2006; Cull and Massie, 1995; Pritchett and Garg, 1978; Keller, 1994; Rapti-Caputo and Bianca, 2008; Valero-Reyes, 2001). In the Basin of Mexico, Michoacan, Guanajuato, Queretaro, Aguascalientes and San Luis Potosi, the existing faults and fractures have directly affected the existing urban infrastructure such as the water supply network and pipe lines. The aforementioned cities rely heavily on groundwater as their main water supply source and an adequate methodology to reflect the risk that faults and fractures have on the aquifer system is needed as they can represent preferential pathways to contaminants located on the surface (i. e. agrochemicals, hydrocarbons).

Study Area

The aquifer of Celaya is located in the La Laja Valley, in the central part of the State of Guanajuato in central Mexico (Fig. 1). The vulnerability of this aquifer has increased due to a series of faults recently detected (Trujillo-Candelaria, 1999; Valero-Reyes, 2001). A preliminary analysis of the region shows that a tectonic component is currently affecting this area, increasing the vulnerability of the area's only water supply source.

This vulnerability increase is found throughout the state of Guanajuato as the valleys of Los Apaseos and Salamanca are affected by the appearance of faults and fractures which have been associated to land subsidence in other areas but which may be caused by regional tectonics in this region, as has been found in different studies (Charles and Kreitler, 2006; Cull and Massie, 1995; Pritchett and Garg, 1978; Keller, 1994; Rapti-Caputo and Bianca, 2008).

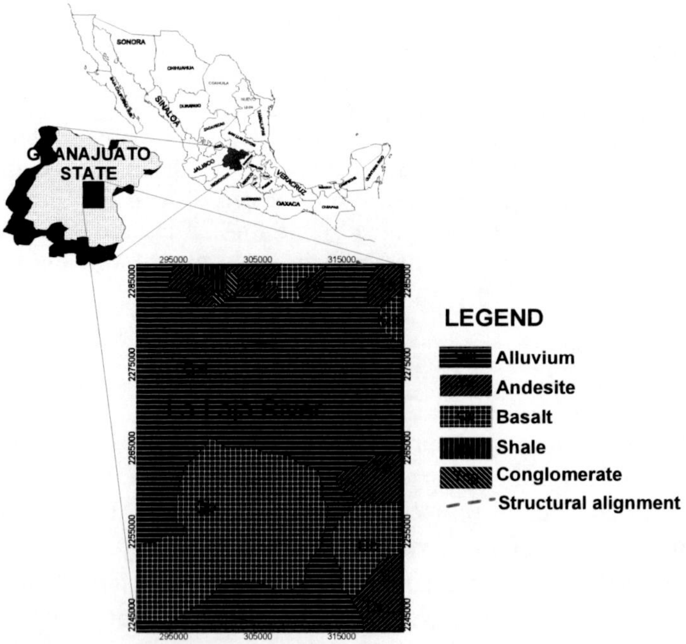

Figure 1. La Laja Valley.

Geology

Most of the outcropping rocks in the study area are volcanic rocks from the Neovolcanic range and are mainly comprised of basalts, andesites and dacites (Demant, 1978). The sedimentary rocks found in this region consist of an alternate sequence of limestones and lutites, which originated from the Mexican Eastern mountain range known as Sierra Madre Oriental. These rocks outcrop in a small region near Juventino Rosas and north of San Miguel de Allende near the northern limit of the Neovolcanic range with the Central Plateau. Acidic rocks, mainly rhyiolites from eruptions of the Western mountain range (Sierra Madre Occidental) outcrop on the northern and northeastern regions of the study area (Fig. 1).

A thick layer of alluvial deposits comprised of fluvial sediments from La Laja river fills the tectonic trenches found in this region, ranging from clays to gravel. As these sediments originate from the Pliocene, different layers of tuff and basalt are found interbedded within them, created during the representative volcanic activity of the Plioquaternary.

The faults that are currently affecting the city of Celaya show a N-NW orientation with an approximate length of 8 km, with an "echelon" structural pattern exhibiting left-sided compresive stresses (Fig. 1). These faults continue along their preferential direction; however, it is only within the city of Celaya that traces of vertical displacements are found.

Methodology

A large number of methods to estimate aquifer vulnerability have been applied on environmental impact studies (Crema et al., 1998; Foster and Hirata, 1991; Hirata and Reboucas, 1999; Gogu and Dessargues, 2000; Gustafson, 1989; Haertlé, 1983; Silva-Garcia and Ramos-Leal, 1998). However, geological features such as faults and fractures of tectonic origin are not considered by most of these methods (Silva-Garcia and Ramos-Leal 1998; Rodríguez-Castillo et al., 1999; Borja and Rodríguez, 2005) and aquifer vulnerability analyses are needed in regions with tectonic activity such as the central part of Mexico. Accordingly, a comparison of different methodologies is undertaken, in order to find out which methodology better reflects the effect that structures of tectonic origin have on the vulnerability of an aquifer (Fig. 1).

a) Sintacs Method

This methodology was developed by Civita and De Maio, (1997) and its name is derived from the seven parameters that are evaluated (Table 1), and to which a value that ranges from 1 to 10 is assigned as detailed below (Figs. 2a, 2b, 2c, 2d). Parameters like the depth of depth water, *net recharge*, hydraulic conductivity and topographical slope can be obtained using functions polynomials (Ramos-Leal, 2005; Civita and De Maio, 1997).

Table 1. Parameters and associated weights of the SINTACS method (from Civita and De Maio, 1997).

	Parameters	Normal weight w_i
S	**S**oggiacenza (*depth to water*)	5
I	**I**nfiltracione efficace (*net recharge*)	4
N	**N**on-saturo –efecto di autodepurazione del- (*impact of vadose zone*)	5
T	**T**ipología della copertura (*type of soil cover*)	4
A	**A**cquifero –caratteristiche idrogeologiche del- (*lithology of the aquifer*)	3
C	**C**onducitibilitá hidráulica dell' acquifero (*hydraulic conductivity of the aquifer*)	3
S	**S**uperficie topográfica –acclivitá della- (*slope*)	2

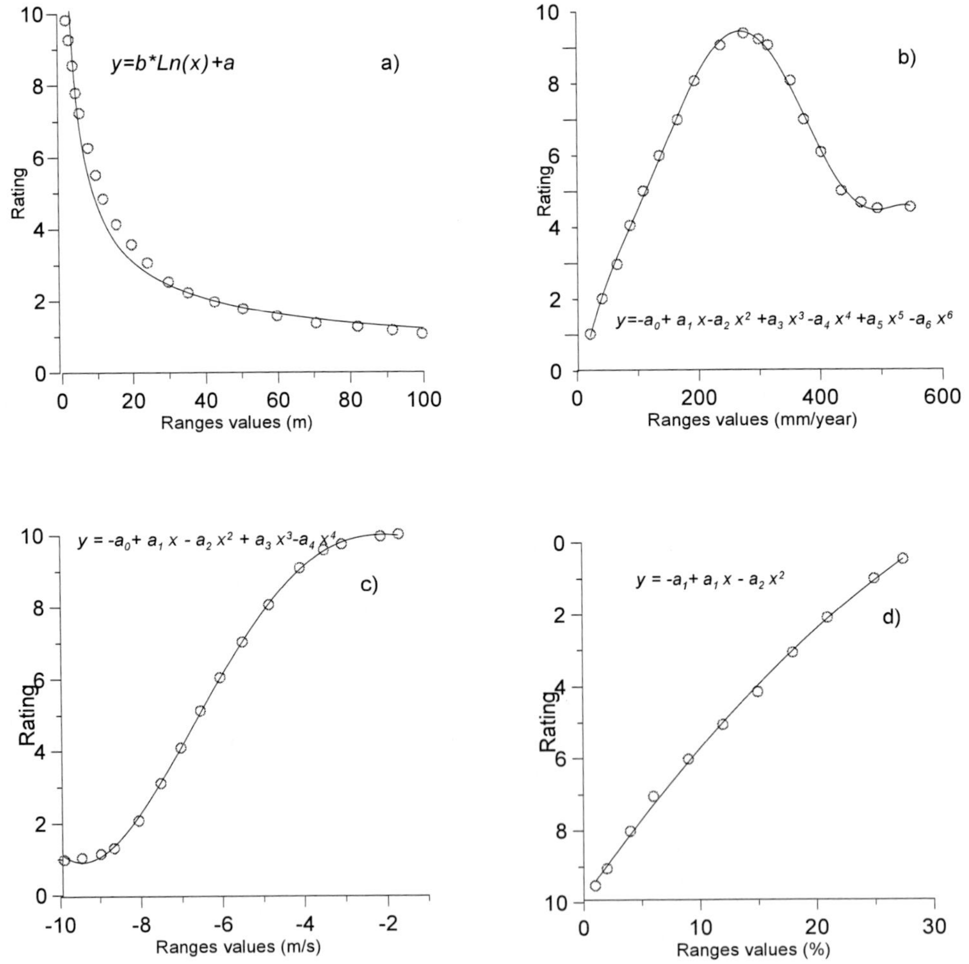

Figure 2. Value assignment to different parameters in the SINTAC methodology: a) depth to water value, b) net recharge, c) hydraulic conductivity, d) topographical gradient.

Depth to Groundwater (S)

A depth between 0-100 m is considered and reflects the thickness of the material through which a contaminant has to travel before reaching the aquifer (Fig. 2a). As in Mexico, depths of more than 150 m are found, the value assigned to this parameter was determined through a logarithmic regression (Ramos-Leal, 2005) which perfectly fits the depth-value relationship of Civita and De Maio, 1997).

Net Recharge (I)

This parameter is determined by substracting real evapotranspiration to precipitation, obtaining what is known as net recharge (Aller et al., 1985; Civita and De Maio, 1997).

$$Rn = P\text{-}Er \qquad \text{(mm/year)}$$

where Rn is net recharge (mm), P is mean annual precipitation (mm) and Er is real annual evapotranspiration (mm). This last term can be determined through the Turc equation (Civita and De Maio, 1997) which requires temperature and precipitation:

$$Er = \frac{P}{\sqrt{0.9 + \left(\frac{P^2}{L^2}\right)}} \qquad \text{(mm/year)}$$

where L is a function of temperature (T) and determined by:

$$L = 300 + 25{*}T_c + 0.05{*}T_c^3$$

where Tc represents mean annual temperature and obtained by (Aller et al., 1985; Civita and De Maio, 1997):

$$T_c = \frac{\sum Pm * Tm}{\sum Pm}$$

where Pm is mean monthly precipitation (mm) and Tm mean monthly temperature (°C).

Finally, the value assigned to the Net Recharge parameter is determined by applying the relationship of Fig. 4(b) (Ramos-Leal, 2005; Civita and De Maio, 1997):

Autodepuration Effect of the Non-Saturated Zone (N)

This parameter makes use of lithologic sections. Accordingly in the study area the highest values are found on the recharge zones, due to the presence of fractured basalts. High values are also associated to rhyiolites, volcanic tuffs and conglomerates while the lowest values are found on the valleys, where sediments with large amounts of clay are found (Fig. 3).

Soil type (T)

This parameter is determined in a similar fashion as N. High values are assigned to regions that are covered by permeable soils, while low values are assigned to impermeable soils. This parameter was assigned based on edaphological maps (INEGI, 1995) and on field trips to the study area, with a similar distribution to N (Fig. 4).

Aquifer type (A)

This parameter is derived from lithological data, along with both geological sections and maps. In the study area, the largest values are assigned to fractured volcanic rock, while the lowest are assigned to the valleys due to the presence of low permeability material.

Hydraulic Conductivity (C)

This value is a function of the hydraulic conductivity values of the different hydrogeological units. The largest values in the study area are assigned to fractured volcanic rocks, while the lowest values are assigned to the valley bottom due to the presence of low conductivity material. The K values used were obtained from Guanajuato's Water Comission (CEAG, 1998) and the final value assigned to this parameter was determined through a polynomial equation (Fig. 2(c); Ramos-Leal, 2005).

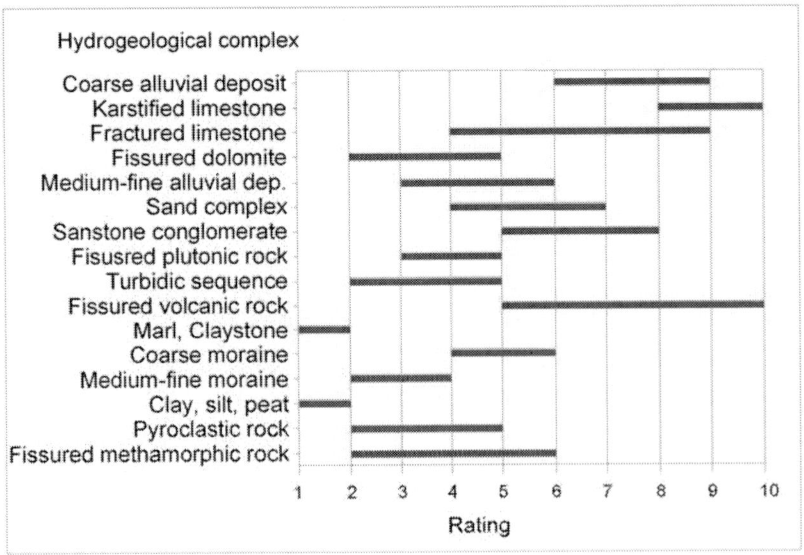

Figure 3. Value assignment for vadose zone (Civita and De Maio, 1997).

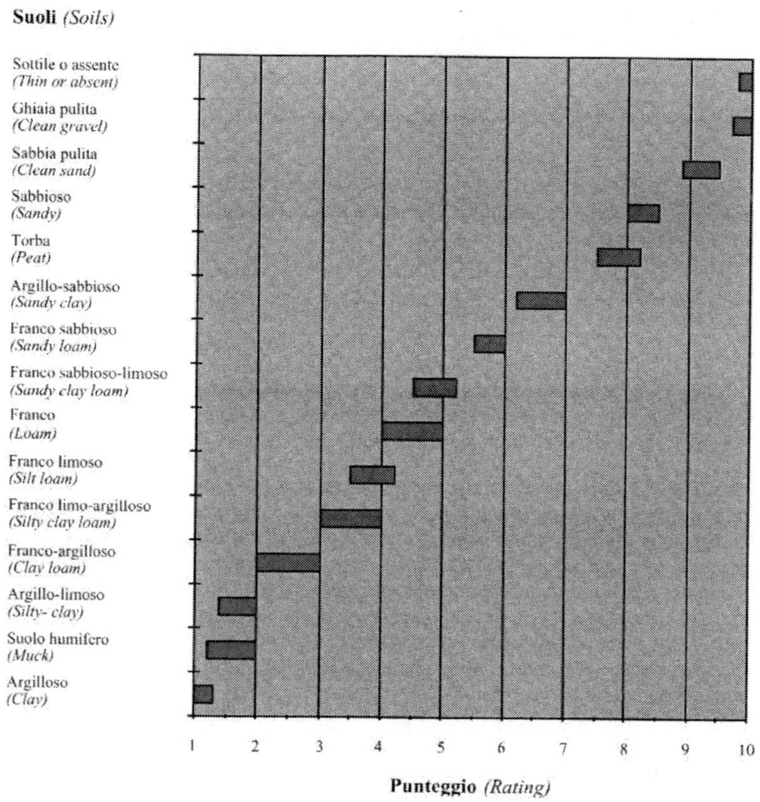

Figure 4. Value assignment for types of soils (Civita and De Maio, 1997).

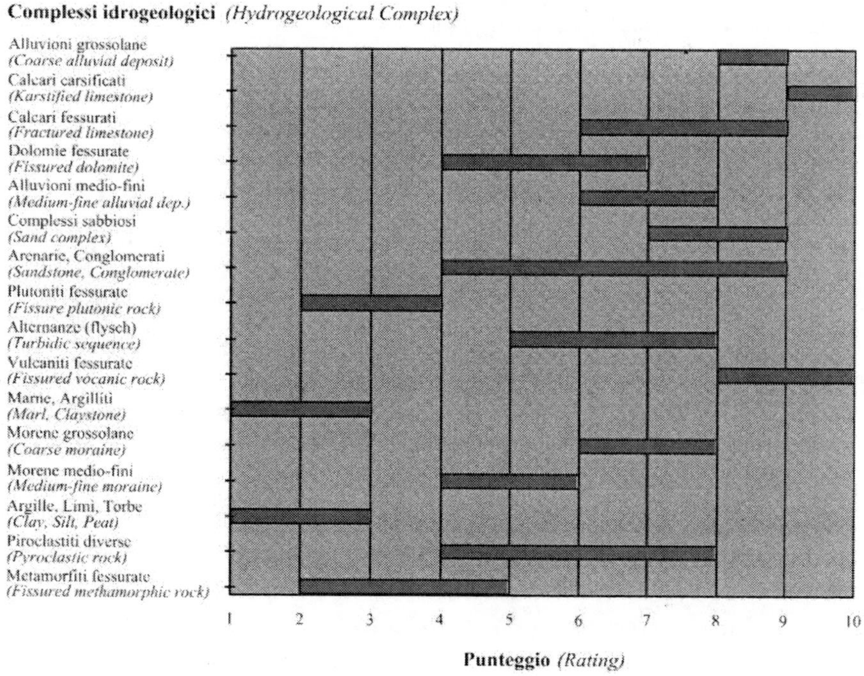

Figure 5. Value assignment for type of aquifer (Civita and De Maio, 1997).

Topographic Gradient (S)

This parameter is determined through slope analysis; for steeper slopes, lower parameter values are assigned as shown in Fig. 2d and was determined from topographic data (INEGI, 1988)

The aforementioned parameters are aggregated into the SINTAC index through the assignment of different weights as follows:

$$S_rS_W + I_rI_W + N_rN_W + T_r\,T_W + A_rA_W + C_rC_W + S_rS_W = I_{\,SINTACS}$$

b) AVI method

This is one of the easiest and fastest methods to determine the vulnerability of an aquifer as only both the hydraulic conductivity and thickness of each stratum overlying the saturated zone are needed. This methodology is based on the resistance $\boldsymbol{c}$ to vertical flow of each material that comprises the vadose zone (Van Stempvoort et al., 1994 and determined by:

$$c = \Sigma bi/Ki$$

where bi represents the thickness of each i stratum, and Ki the hydraulic conductivity of each stratum.

The AVI index represents the total hydraulic resistance of the vadose zone and indicates the time per unitary gradient that it would take for a contaminant to arrive to the saturated zone. A larger c value represents less vulnerability. Due to the large variability of c within the wells analyzed, log(c) is normally used (in years).

C) Radon-222

Due to the fact that faults and fractures modify the permeability of a porous medium, Rn-222 was used in the study area, as this gas dissipates faster on permeable media (Valero-Reyes, 2001). A total of 30 Rn-222 detectors were distributed within the Celaya urban zone and in La Laja valley, where the existence of faults that exhibit vertical displacements of up to 3m and lenghts of 10 km have damaged the urban infrastructure and are considered to affect the aquifer's vulnerability.

d) Water Quality Index

According to Couillard and Lefebre (1985) a water quality index is an algorithm that represents a qualitative measure of water. These indexes provide a global value on water conditions through the ponderation of individual values assigned to physical, chemical o biological parameters which can be measured in the field or in a laboratory (Table 2).

This index takes into account the maximum number of available components and can be obtained from a water quality analysis (i.e. ions, pH, temperature and electric conductivity). The larger the number of components used the better obtained index.

All parameters are ponderated according to their polluting capability, thus an index *Ix* is obtained for each parameter, which are then averaged to obtain an overall *WQI,* according to the following relationship (Conesa, 1993):

$$WQI = k\frac{\sum C_i P_i}{\sum P_i}$$

where Ci is the percentual value of each parameter, Pi represents the weight assigned to each paramenter and k is a value obtained from Table 3, and is a function of the water physical characteristics.

Table 2. Weighting constants as a function of water pollution (physical, chemical and biological characteristics).

Ratin	Aspect	pH	Temp	Hard	EC	OD	Reduc. of pe	Coliformes	N-amoniacal	Detergents	S. Dis.
%	(Subjetivo)		(°C)	(mg-CO3Ca/l)	(Mmhos/cm)	(mg/l)	(mg/l)	(no./100ml)	(mg/l)	(mg/l)	(mg/l)
0	Pésimo	1/14	>50/>-8	1500	16000	0	15	14000	1.25	3	20000
10	Muy-malo	2/13	45/-6	1000	12000	1	12	10000	1	2	10000
20	Malo	3/12	40-4	800	8000	2	10	7000	0.75	1.5	5000
30	Desagradabl	4/11	36/-2	600	5000	3	8	5000	0.5	1	3000
40	Impropio	5/10	32/0	500	3000	3.5	6	4000	0.4	0.75	2000
50	Normal	6/9.5	30/5	400	2500	4	5	3000	0.3	0.5	1500
60	Aceptable	6.5	28/10	300	2000	5	4	2000	0.2	0.25	1000
70	Agradable	9	26/12	200	1500	6	3	1500	0.1	0.1	750
80	Bueno	8.5	24/14	100	1250	6.5	2	1000	0.05	0.06	500
90	Muy-bueno	8	22/15	50	1000	7	1	500	0.03	0.02	250
100	Excelente	7	21/16	25	750	7.5	0.5	50	0	0	100

Rating	Cl	Plaguisidas	Grasa-aceites	SO4	NO3	Cianuros	Na	Ca	Mg	PO2	NO2	DBO
%	(mg/l)	(mg/l)	ppm	(mg/l)	(mg/l)	(mg/l)	(mg/l)	(mg/l)	(mg/l)	(mg/l)	(mg/l)	(mg/l)
0	1500	2	3	1500	100	1	500	1000	500	500	1	15
10	1000	1	2	1000	50	0.6	300	600	300	300	0.5	12
20	700	0.4	1	600	20	0.5	250	500	250	200	0.25	10
30	500	0.2	0.06	400	15	0.4	200	400	200	100	0.2	8
40	300	0.1	0.3	250	10	0.3	150	300	150	50	0.15	6
50	200	0.05	0.15	150	8	0.2	100	200	100	30	0.1	5
60	150	0.025	0.08	100	6	0.1	75	100	75	20	0.05	4
70	100	0.01	0.04	75	4	0.05	50	50	50	10	0.025	3
80	50	0.005	0.02	50	2	0.02	25	25	25	5	0.01	2
90	25	0.001	0.01	25	1	0.01	15	15	15	1	0.005	1
100	0	0	0	0	0	0	10	10	10	0	0	0.5

The percentual value Ci is determined from Table 3.

The most frequently used parameters are ions, Biochemical Oxygen Demand (DBO), dissolved and suspended solids, along with Nitrogen, Phosphorous, Sulfur, pH, hardness, turbidity, electric conductivity and both toxic and pathogen elements. In the present study, only major ions were used (CEASG, 1998).

Table 3. Percentual value for each parameter.

k	Water characteristics
1.0	Clear water, without apparent pollution
0.75	Water with Light color, foam, non-natural turbidity
0.50	Apparently polluted water with strong odor.
0.25	Waste waters with fermentation and strong odor.

The obtained WQI value ranges from 0 to 100, which is associated to water pollution according to Table 4:

Table 4. Water quality index (WQI) and its associated pollution level.

WQI VALUE	WATER CLASSIFICATION
0 – 30	Highly contaminated
30 – 50	Contaminated
50 – 70	Low contamination
70 – 85	Acceptable
85 – 100	Excellent

It should be noted that although given water can be classified as water of excellent quality, other parameters should be considered. If an excellent value is obtained (i.e. WQI=90) and the total fecal coliform value exceeds the maximum allowed value, the index would need to be modified. The WQI index is a valuable tool to estimate the complexity involved in the multiple uses of water.

E) Contamination Index (CI)

Another way to estimate groundwater pollution is through the use of the Contamination Index developed by Backman et al. (1998) which represents the summation of the components that exceed the maximum concentration values (NOM-127, 1994).

The contamination level Cd was obtained for each sample, using the CEASG (1998) database, which comprises a total of 30 samples from which the major ions were determined with an analytical error below 10%. This index was determined by applying the following relationship:

$$C_d = \sum_{i=1}^{n} C_{fi}$$

where Cfi if the contamination factor for the i-th component obtained through:

$$C_{fi} = \frac{C_{Ai}}{C_{Ni}} - 1$$

where C_{Ai} is the analytical value of the i-th component and C_{Ni} is the maximum allowed concentration for the same i-th component.

In order to apply this methodology in Mexico, the maximum allowed values are determined from the Mexican Norm NOM-127-SSA1-1994: "Environmental health, water for human use and consumption. Maximum allowed quality values and water treatment methodologies for water potabilization".

Results and Discussion

a) SINTACS

The vulnerability values obtained for water depth range from 2 in the mountainous region to 1 in the valley zone (Figs 2 and 6), while the maximum values for the Net Recharge parameter (6) are located in the SE region of the study area, with the lowest Net Recharge values (1) found in the valley, in the area where faults are found (Figs 3 and 6). The values assigned to the vadose zone are high (7-6) on both the northern and southern parts of the valley, while values between 2 and 5 are found in the central part of the valley (Figs. 4 and 6), with a similar distribution as that of the Soil Type vulnerability value (Figs. 5 and 6).

The value assigned to aquifer type vulnerability is large throughout the study area (Figs. 6 and 9), although the lowest values are found in the eastern area, while the Hydraulic Conductivity value is large on the southern region, with the lowest values found on the north-central region. Finally, the topographic index is high on the valley and low on the hills at the valley's southern region, as expected.

Through the aggregation of the aforementioned indexes, a normalized vulnerability map for the study area was developed (Fig. 9). The SINTACS map shows large values of 70-80 in both the southern and northern regions, while the lowest values (10-30) are found within the valley.

The Celaya urban zone, where fractures are present reaches a value of 30, thus no zoning associated with these structures is shown in the SINTACS map (Fig. 7), even though they were digitized and assigned a large ponderation value.

b) AVI

The AVI map shows low or negative values on the southern region of the valley, indicating the most vulnerable zones. The largest values, are located on the eastern part of the study area, indicating the least vulnerable zones. In this case, the fault zone shows a larger vulnerability (Fig. 8) than the one obtained with SINTAC.

c) Rn-222

This method clearly reflects the vulnerability associated with the fault zone, showing anomalies along this zone. The obtained values were normalized to a vulnerability range of 0-1, where 1 represents the lowest Rn-222 values and 0 the highest. The results obtained with this methodology were corroborated to the physicochemical anomalies measured in the valley as in areas classified as highly vulnerable, abnormal pH and temperature values were measured, pointing to a relationship between these values and the existence of faults (Fig. 9).

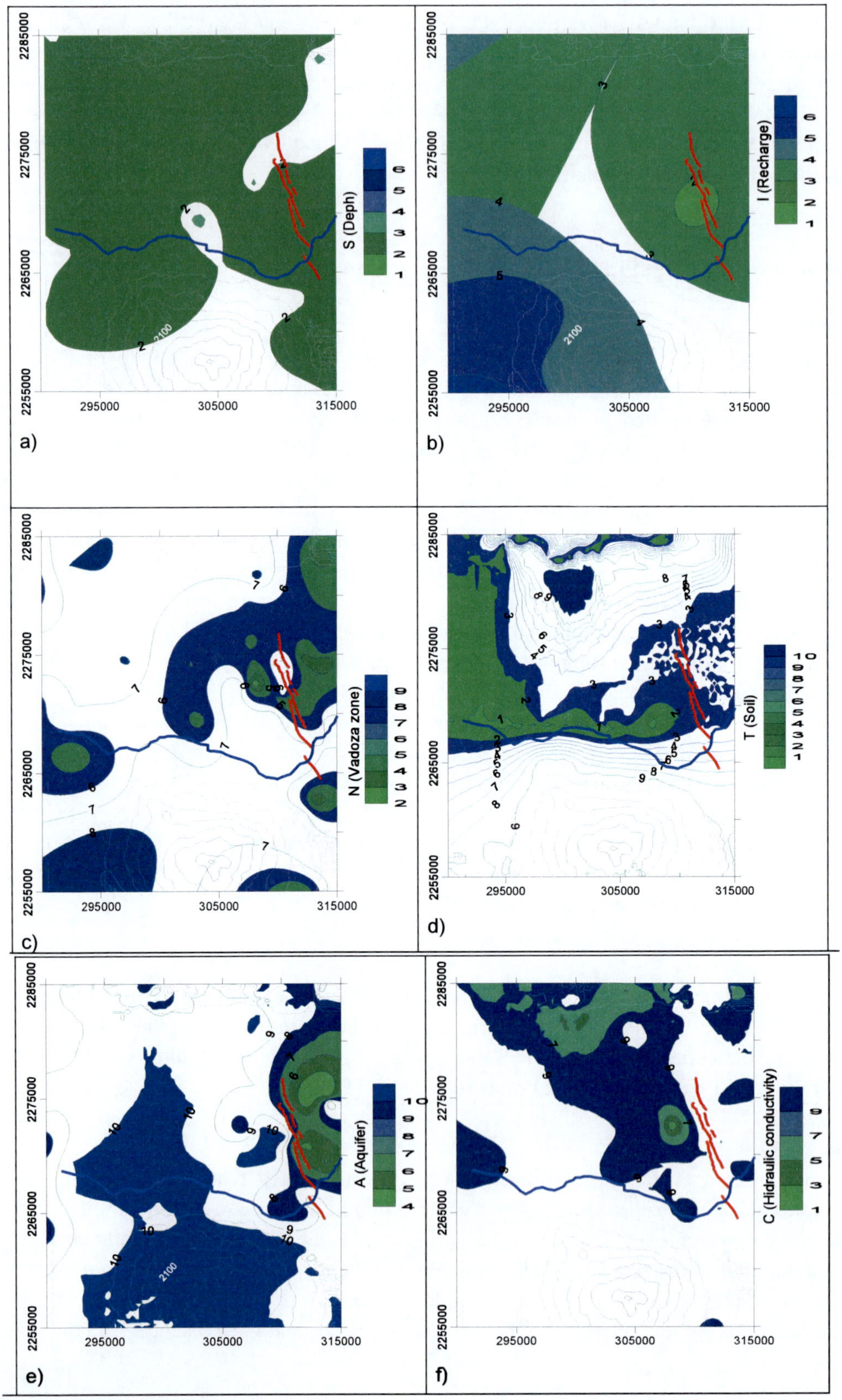
2285000
2275000
2265000
2255000
295000
305000
315000
S (Deph)
I (Recharge)
N (Vadoza zone)
T (Soil)
A (Aquifer)
C (Hidraulic conductivity)
2100
a)
b)
c)
d)
e)
f)

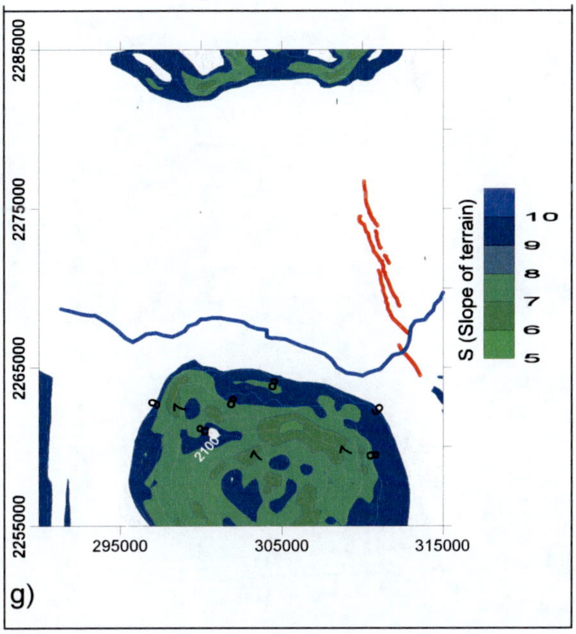

Figure 6. Distribution of the parameters used in the SINTACS methodology a) depth to water, b) net recharge, c) vadose zone, d) soil type, e) aquifer type, f) hydraulic conductivity, and g) topographic gradient.

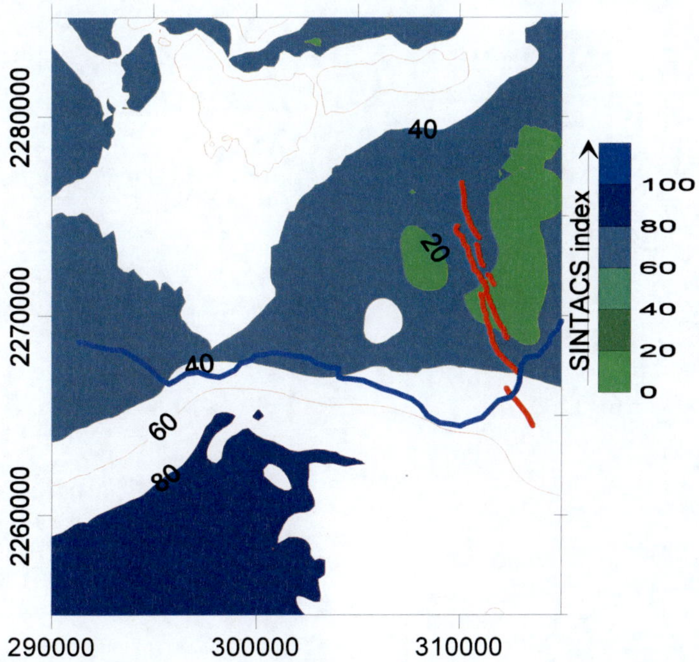

Figure 7. Vulnerability map of SINTACS for the Celaya aquifer.

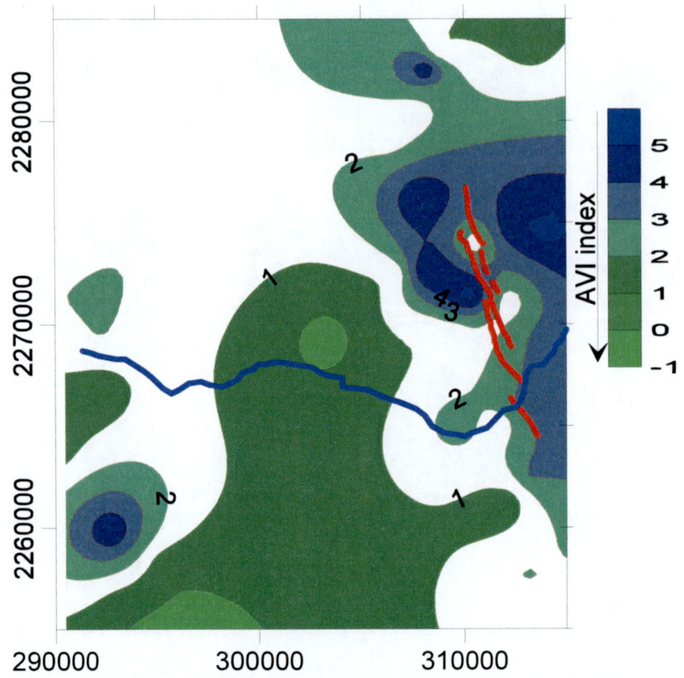

Figure 8. Vulnerability map of AVI for the Celaya aquifer.

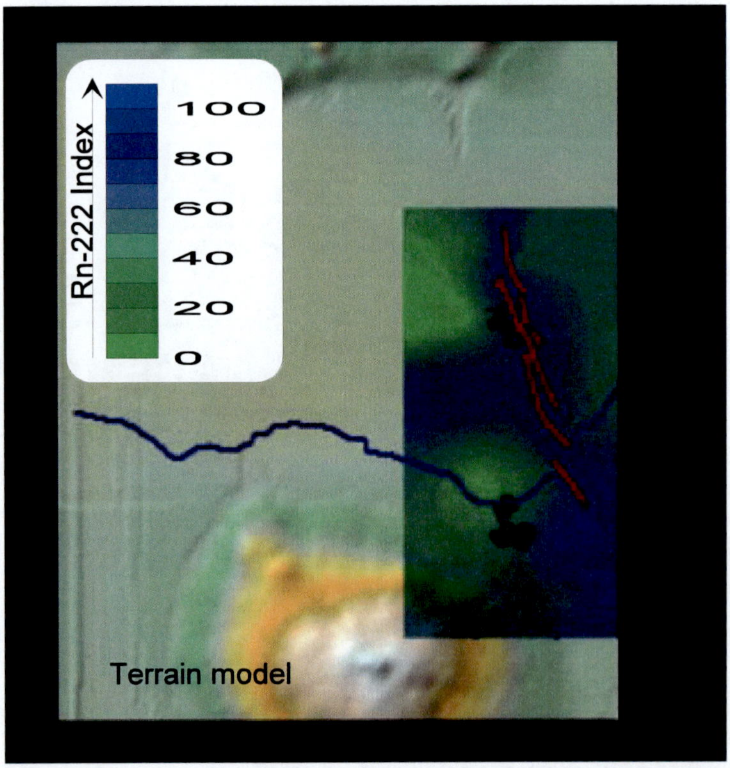

Figures 9. Vulnerability map based on Rn-222 (Rn-222 was applied in a portion of the valley, modified from Valero-Reyes, 2001).

Water Quality (WQI) and Water Pollution (CI) Indexes

The spatial distribution of the WQI index shows that water of better quality is found in the central part of the valley, while the lowest-quality water is found both at the valley's eastern and western boundaries (Fig. 10(a)). This distribution is also observed on the CI index (Fig. 10(b)) and is related to the geologic faults located in the western part of the valley.

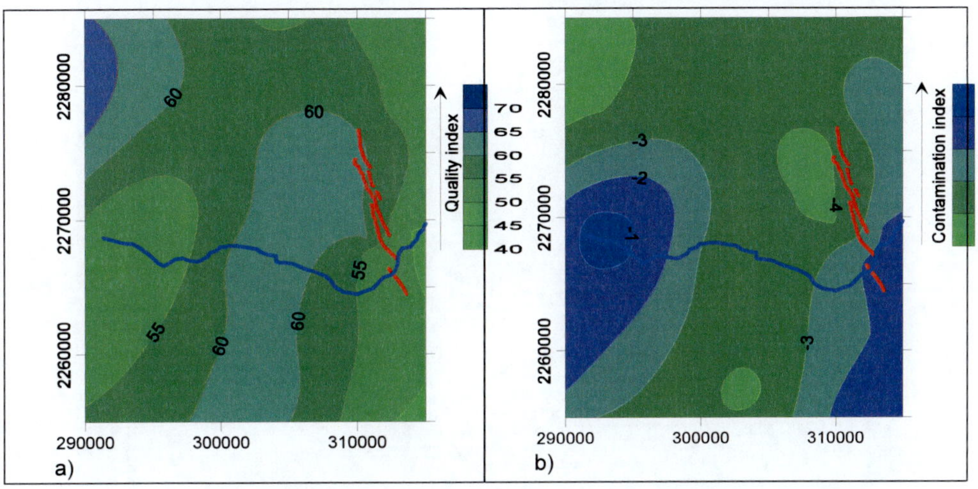

Figure 10. Water quality indexes for the Celaya aquifer: (a) Water Quality Index and (b) Water Pollution Index.

CONCLUSIONS

Through the evaluation of different vulnerability indexes (SINTACS, AVI, WQI, CI, Rn-222) in a tectonically active region, it was found that traditional methods (i.e. SINTACS and AVI) do not reflect the effect that faults and fractures have on an aquifer's vulnerability to become polluted (Fig. 11). This effect was reflected by the isotopic methodology used in this work (Rn-222) which can also be applied in other urban centres of central Mexico that heavily rely on groundwater as their main water supply source and where the presence of faults and fractures may work as preferential pathways to surface contaminants. In addition, the vulnerability map obtained with the Rn-222 method showed the presence of faults that had not been previously detected, making it a valuable tool in vulnerability assessments.

Another aspect that also plays an important role in the appearance of fractures is land subsidence caused by heavy extraction of groundwater. Thus, aquifers that were thought of as non-vulnerable may change its conditions through the appearance of fractures on low permeability sediments.

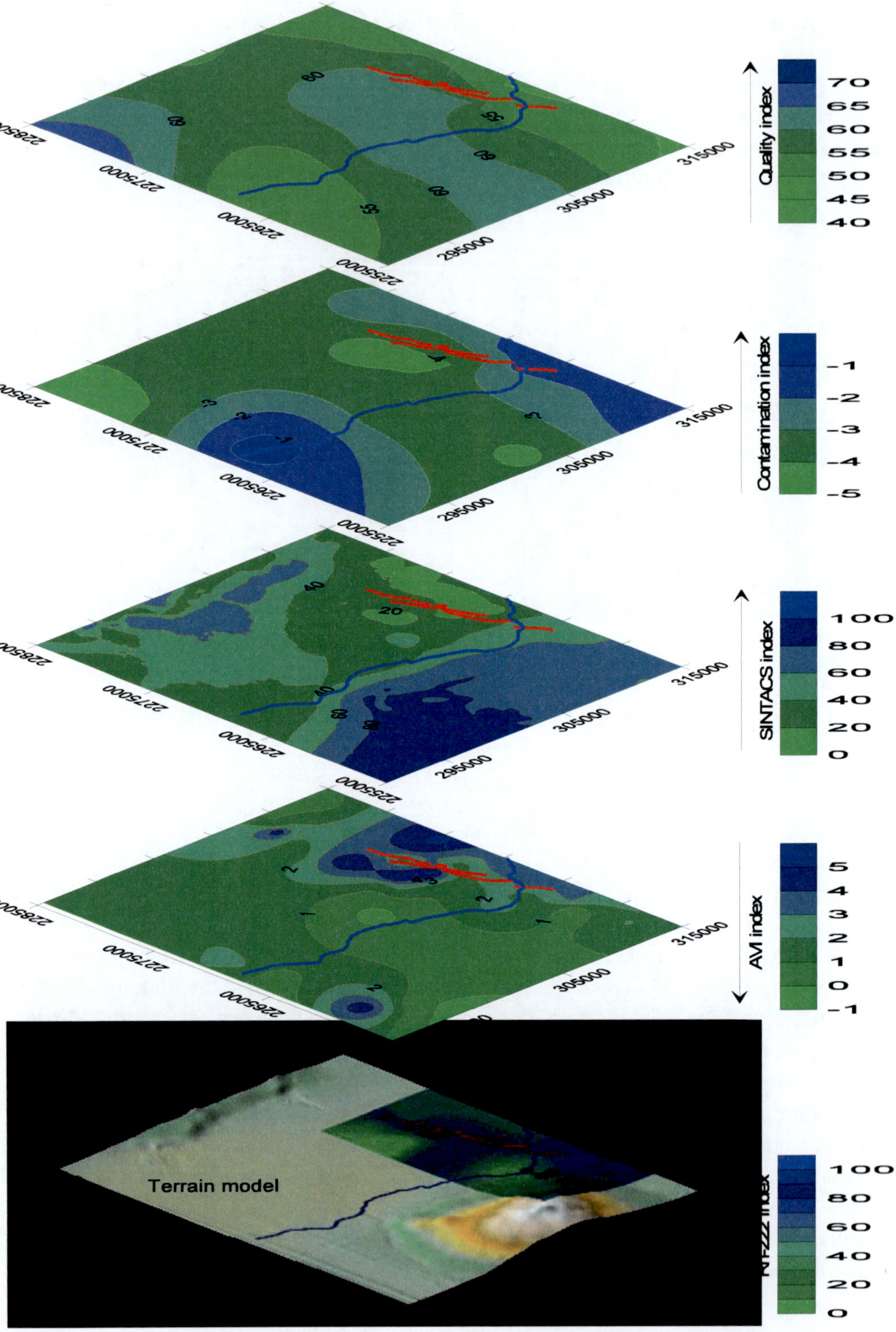

Figure 11. Comparison of the vulnerability indexes for the aquifer of Celaya.

REFERENCES

[1] Aller, L. T. Bennet, J. H. Lehr, R. J. Petty, and G. Hackett, 1985. DRASTIC; A Standard System for Evaluation Groundwater Pollution Using Hydrogeologic Setting, EPA/600/2-85/018 US EPA.

[2] Backman, B., Bodis, D., Lahermo, P., Rapant, S. and Tarvainen, T. 1998. Appplication of a Groundwater Contamination Index in Finland and Slovakia. *Enviromental Geology*, 36 (1-2), 55-64 pp.

[3] Borja O., R. I. and Rodríguez C., R. Aquifer vulnerability changes due to faults and riverbeds in Salamanca, Guanajuato, Mexico. Geofísica Internacional (2004), Vol. 43, Num. 4, pp. 623-628.

[1] Charles W. and Kreitler, C.W. 2006. Fault Control of Subsidence, Houston, Texas *Ground Water.* Volume 15 Issue 3, 203-214 pp.

[4] Civita M. y and De Maio, 1997. SINTACS. Un Sistema Parametrico per la Valutazione e la Cartografia della Vulnerabilitá Degli Acquiferi All'inquinamento. Metodologia & Automatizzazione. Pitagora Editrice Bologna. pp 191.

[5] Conesa, F., 1993. Guía Metodológica para la Evaluación del Impacto Ambiental. Ediciones Mundi Prensa. pp. 276.

[6] Couillard, D. and Lefebvre, Y. 1985. Analysis of Water Quality Indices. *Journal of Environmental Management.* 21: 161-179.

[7] Comisión Estatal de Aguas y Saneamiento de Guanajuato (CEASG), 1998. Estudio Hidrogeológico y modelo matemático del acuífero del valle de Los Apaseos, Guanajuato. Reporte interno, 225 pp.

[8] Crema, G., Milli, M., Napoleone, G. and Tiberi, P. 1998. Vertical Vulnerabilty Evaluation of Middle-low Esino River valley Alluvial Aquifer (Marche, Italy). *Enviromental Geology*. 34 (2/3). 88-94 pp.

[9] Cull, J.P. and Massie, D. 1995. Aquifer mapping and groundwater quality in faults and fractures. Exploration Geophysics. 33(2) 122 – 126 pp.

[10] Demant, A. 1978. Características del eje Neovolcánico Transmexicano y sus Problemas de Interpretación, UNAM, Inst. Geología Revista, vol. 2, no. 2, p.172-187.

[11] Foster, S. and Hirata, R. 1991. "Determinación del Riesgo de Contaminación de Aguas Subterráneas". Centro Panamericano de Ingeniería Sanitaria y Ciencias del Ambiente (CEPIS), Lima, Perú, 81 pp.

[12] Gogu, R. C. and Dessargues, A., 2000. Current Trend and Future Changes in Groundwater Vulnerability Assessment Using Overlay and Index Methods. *Environmental Geology,* 39 (36), Abril.

[13] Gustafson D. I., 1989. Groundwater Ubiquity Score: A Simple Method for Assessing Pesticide Leachability. *Environmental Toxicology and Chemistry* V. 8 pp 337-357.

[14] Haertlé, T. 1983. Method of Working and Employment Of EDP During The Preparation Of Groundwater Vulnerability Maps. In Groundwater In Water Resources Planning, UNESCO. INTER. SYMP, Proc. Koblenz, Alemania, 1983, UNESCO/IAH/IAHS. Vol. II, 1073-1085 pp.

[15] Hirata, R. and Reboucas, A., 1999. La Protección de los Recursos Hídricos Subterráneos: Una Visión Integrada Basada en Perímetros de Protección y

Vulnerabilidad de Acuíferos. Boletín Geologico Minero. Vol. 110-4 Ano 1999 (423-436).

[16] INEGI, 1995. Carta edafológica, 1:50000 F14C64 (Celaya).

[17] INEGI, 1988. Carta topográfica 1:50000 F14C64 (Celaya).

[18] Keller, B. 1994. Water Characterization for Enviromental Hydrogeologic Monitoring in Fault Zones: Examples from California, Potencial Aplications in Chile. *Journal of Suth American Earth Science*, V. 7. Nos. 3/4. 295-300.

[19] Norma Oficial Mexicana NOM-127-SSA1-1994, "Salud ambiental, agua para uso y consumo humano-limites permisibles de calidad y tratamientos a que debe someterse el agua para su potabilización".

[20] Pritchett, J. W. and Garg, S. K. 1978. Flow in an aquifer charged with hot water from a fault zone. *Pure and Applied Geophysics*. Volume 117, Numbers 1-2 / enero de 1978, 1420-9136 pp.

[21] Ramos Leal, J. A. 1998a. "Análisis del Papel Contaminante del Área del Panteón Metropolitano y del Basurero de San Lorenzo". Trabajo de Investigación, Posgrado en Ciencias, UNAM-Inst. Geofis. Inédito.

[22] Ramos Leal, J. A. 1998b, "Diagnóstico de Impacto Ambiental en Sitios con Fuentes Contaminantes Potenciales o Activas en la Cuenca de México", Memorias. 1er. Seminario del Posgrado en Ciencias de la Tierra, Instituto de Geofísica, UNAM.

[23] Ramos Leal, J. A. 2005. Curso De Vulnerabilidad Acuifera e Indices De Calidad Del Agua. IPICYT, Reporte Interno. 63 p.

[24] Rapti-Caputo, D. and Bianca, M. 2008, Carbonate aquifer systems: fault-controlled groundwater flow and vulnerability. *Geophysical Research.* Vol. 10, EGU2008-A-07401.

[25] Rodríguez-Castillo, R., Ramos-Leal, J., A. and Mejía, J.A. 1999. Geological and Hydrogeological Controls Over Organic Compounds Groundwater Pollution. Its Implication in the Urban Water Supply of Salamanca, Gto. México. Proceeding. AGU. San Fransisco, California.

[26] Silva-García, J.T. and Ramos-Leal, J.A., 1998, "Elementos de Vulnerabilidad de Acuíferos, para la Protección de Manantiales. Caso Lago de Camecuaro, Michoacán", Memorias. Unión Geofísica Mexicana, Congreso Anual, Puerto Vallarta, Jalisco, México.

[27] Trujillo- Candelaria, 1999. Fallamiento De Terrenos por Sobreexplotación de Acuíferos, Caso Celaya. Memorias. Seminario Estatal de Investigación Sobre el Agua; Desafío del Agua en Guanajuato hacia el 2000, Instituto de Geofísica, UNAM; Cd. de León, Gto.

[28] Turc, L., 1955. Le Bilan D´Eau Des Sols. Relations Entre Les Precipitations, L´Evaporation Et L´Ecoulement. Ann. Agro. 5-131.

[29] Valero- Reyes, A., 2001. " Riesgo Geológico en la Ciudad de Celaya, Guanajuato, Utilizando la Detección de Radón-222.". ESIA-iPN, Tesis de Maestria. Inédita.

[30] Van Stempvoort, D., L., Ewert and L. Wassenaar, 1995. AVI: A method for Groundwater Protection Mapping in the Praire Province of Canada. PPWB Report No. 114, National Hydrology Research Institute, Saskatoon Saskatchewan, Canada.

[31] Vrba, J., and Zaporozec, A., 1994. Guidebook on Mapping Groundwater Vulnerability. *International Association of Hydrogeologists.* Vol. 16, 131 pp.

In: Remediation of Soils and Aquifers
Editors: Luis G. Torres and Erick R. Bandala
ISBN 978-1-60741-372-1

Chapter 12

CLOSURE OF TAILINGS DAMS AND REMEDIATION OF SOILS CONTAMINATED WITH MINING WASTES

José E. Santos Jallath
Facultad de Ingeniería-UNAM

1. INTRODUCTION

Historically, one of the most important economic activities of Mexico has been mining. For over 500 years, mineral fields with diverse metal contents have been exploited in the whole territory. During this long time, the processes used to concentrate those metals of economic value have evolved considerably, from the ancient mercury amalgamation process to extract gold and silver, the revolutionary flotation process for sulfide minerals, the cyanide process to recover precious metals, and, more recently, the lixiviation processes (leaching) that allow for the extraction of a large amount of metals in solution.

The wastes generated by these processes are known as tailings (in Mexico, they are known as "jales"[1]) and are, generally, stored in deposits close to the exploitation zones, these deposits have been termed tailings dams (CNA, 2000; AIMMGM, 1993, Environment Australia, 1995). These wastes are characterized by the content of a large variety of minerals, mainly silicates, oxides, hydroxides, carbonates, sulfides, and sulfates, aside from low concentrations of some metals and metalloids that, because of their toxicity, can represent a risk for the environment, among these are arsenic, lead, cadmium, mercury, selenium, and others (Lottermoser, 2007).

Large volumes of tailings, deposited for so many years in the mining zones, have been exposed to meteorological phenomena, such as solar radiation, rain, evaporation, and winds, which have contributed to their dispersion and transportation to zones near the deposits, affecting soils, surface waters, and, in some cases, aquifers.

---

[1] Tailings (Jales) are a fine material originated by the concentration processes in which there is a previous grinding stage. Other processes, such as lixiviation or cyanidation, can be applied to coarse granules material that do not go through the grinding process, and in this case the deposit of the waste is called "terrero" in México.

In Mexico, many studies have been performed on the environmental effects produced by the dispersion of tailings in mining zones, such, as Zimapán, state of Hidalgo (Ongley *et al.*, 2007; Romero *et al.*, 2006), Taxco, state of Guerrero (Armienta *et al.*, 2003; Romero *et al.*, 2007), Santa María de la Paz, state of San Luís Potosí (Castro-Larragoitia *et al.*, 1997; Razo *et al.*, 2004), and in Guanajuato, state of Guanajuato (García-Meza *et al.*, 2004; Mendoza-Amézquita *et al.*, 2006), just to name some examples. Those studies have been generally addressed at investigating the physical and chemical processes that occur in the tailings dams once they are not being operated any more[2], at defining the mechanisms of dispersion and release of metals and metalloids and at determining the magnitude and extension of the produced affectation. Detailed studies have been made on the oxidation process occurring in tailings and on the generation of acidity, these processes have as end product the release of metals and metalloids in solution, representing a risk for the environment and the health of nearby populations, as is the case with arsenic, lead, cadmium, nickel, selenium, copper, and zinc, among others.

However, beyond the characterization of tailings and the study of the physicochemical processes occurring in these deposits, it is necessary to study and assess existing alternatives to control their dispersion and reduce their impact on soils, surface waters, aquifers, and the health of populations. The fundamental point in the management of tailings during and after the operation of the dams is to avoid their dispersion and to ensure their stabilization, both physical and chemical; for this, the criteria for the design, construction, and operation of tailings dams are critical aspects, aside from the measures taken for their closure. On the other side, clean-up and remediation of tailings-contaminated soils is quite complex, from the operating and economic point of view, because of the large amounts of land that are affected with this waste and due to the difficult conditions of the terrain that quite often limit the use of the needed equipment and machinery.

All around the world diverse measures to control the dispersion of tailings and to remediate the contaminated soils have been proposed and instrumented (Sengupa, 1993; Ripley, 1996; Lottermoser, 2007); however, in Mexico, the application of some of these measures is limited mainly because of their high cost. This situation makes it necessary to assess practical alternatives that adapt to the technical and economical conditions of the mining industry of the country. For example, in countries like Canada, United States of America, and Australia, these deposits are covered with synthetic materials (geomembranes), or are saturated with water (flooding the deposits) to avoid oxidation and dispersion. Other alternatives have also been used, such as depositing the tailings together with neutralizing minerals to avoid the generation of acidity, or to fill up the deposits with tailings to avoid depositing them on the surface of the terrain (Sengupa, 1993; Lottermoser, 2007, Environment Australia, 1997).

In Mexico, the procedure most used for the closure of tailings dams after their operation consists of covering them with natural materials (soil, ground rocks, and organic matter) followed by revegetation with species native to the mining regions. Although, this procedure does not warrant complete retention of the tailings, it does help to avoid their dispersion.

---

[2] During the operation of tailings dams, the waste is deposited mixed with water forming a pulp. When the dam ceases to operate, the water starts to be lost through filtration and evaporation, and oxidation processes of the metallic sulfides starts once the tailings dry out.

The demonstrated capacity exhibited by plants to assimilate and, in many cases, to retain or accumulate metals and metalloids (US EPA, 2000; Baker *et al.*, 2000; McGrath *et al.*, 2000; Chaney *et al.*, 2000), allow us to assert that an adequate selection of the vegetal species, to be used in the revegetation of a tailings dam, could be very effective to warrant physical stabilization of the deposits to avoid chemical release of metals and metalloids, to control the generation of acidity from the oxidation of sulfide minerals, and to use the plants in remediation processes of the metal-affected soils.

The remediation processes of contaminated soils known as phytoremediation methods are based precisely on the use of plants capable of removing or transforming the contaminants present in soils, by means of diverse processes, mainly absorption, microbial degradation, phase-changing stabilization (Reddi and Inyang, 2000). These features of the plants can be exploited in the closure of tailings dams to avoid the release of metals and metalloids that can represent an environmental risk.

For a mining country such as Mexico, the study of the plants that grow in the diverse mining regions is fundamental so that the closing procedure of tailings dams by covering them with natural matters, followed by revegetation, becomes effective and instrumental in avoiding future environmental damage.

2. Extraction and Processing of Minerals

Extraction of mineral resources is accomplished by means of underground exploitation or in open pits. Underground exploitation is applied to deep ore bodies and special infrastructure must be built to reach the underground's interior and extract the mineral. The particularity of this type of exploitation is that the surface of the terrain is not significantly altered and the depths can reach several hundreds of meters. On the other hand, open pit exploitation is applied to exploit ore bodies close to the surface, and their main characteristic is that the vegetation and surface soil must be removed to reach the mineral, this implies deterioration of large portions of land. Another important difference between both methods is that in open pit exploitation a large mineral volume is produced with respect to the underground exploitation and this is directly related with the volume of produced wastes that will have to be deposited in an area nearby to the mine.

Once the mineral has been extracted from the mine, a concentration process is undertaken aimed at separating the minerals with economic value (ore) from those lacking it (gangue). These concentration methods vary and are based on the physical and chemical properties of the ore mineral and waste rocks to achieve their separation. There are physical methods (magnetic separation, gravimetric separation, among others), physicochemical methods (flotation process), and chemical methods (cyanidation and leaching).

Concentration methods, such as flotation, magnetic and gravimetric separation, include a crushing and grinding stage, which consists of reducing the size of the mineral below 150 μm in most cases. The objective of this size reduction is to release the ore mineral and achieve a better separation, that is, a better recovery of the economically valued elements. These processes yield, as final product, one or several mineral ore concentrates and a waste called tailings (jales)[3]. Tailings are a very fine material and most of this material has a particle size between 150 and 48 μm.

In Mexico, since the beginnings of the last century, the concentration method most widely used has been the flotation process, which is applied mainly to sulfide minerals. In this process, the mineral goes through a stage of grinding with water and is then passed through a series of cells, to which diverse reagents are applied to modify the surface characteristics of the given minerals and allow them to float adhered to air bubbles that form inside the cells. The floating material forms the concentrate, and what is left in the bottom of the cell constitutes the tailings or flotation wastes.

However, currently in Mexico and in other countries, the cyanidation and leaching processes in piles are being increasingly used mainly due to the high prices reached by metals, such as cold, silver, and copper. These processes do not include the grinding stage, the mineral extracted from the mine is placed on "patios", previously conditioned, forming piles. These piles are irrigated with a water-dissolved sodium cyanide solution, in the case of cyanidation, or of sulfuric acid also dissolved in water, in the case of leaching. The solution acts on the mineral dissolving the elements of interest that are recovered by a drainage system placed on the bottom of the pile. Afterwards, the material is sent to a separating process in which the metallic elements are recovered as solids. Once the economically valued elements have been exhausted, the mineral stored in the pile constitutes the final waste of the process, constituting cyanide or lixiviation piles; in Mexico, they are called "lixiviation terrero" or "cyanidation terrero".

Tailings and the lixiviation and cyanide piles constitute the main waste in mineral processing, and they can represent a risk for the environment if the necessary preventive and control measures are not taken to avoid their dispersion, as these wastes will remain for a long time at the storage site.

3. Impacts Caused on the Environment

Mining activity, for its particular characteristics, causes diverse environmental alterations, which are in function of the type of exploitation being performed, of the characteristics inherent to the exploited mineral, and of the environmental conditions of the site where the activity is carried out.

In general terms, open pit mines cause greater affectation in the surface of the terrain because is requires stripping of surface soil to extract the mineral, and as the mine is further deepened it is necessary to widen the opening of the surface to extract the mineral to warrant stability of the slopes. Besides, the area occupied by the facilities and the waste deposits is larger due to the volume of the mineral produced by this type of mines. For example, an open pit mine can process more than 10 000 tons of mineral per day and this implies the generation at least of three times this volume in wastes. In Mexico, the largest open pit mines process daily around 80 000 tons of the corresponding mineral.

Generally, the underground mines reach a lower production as compared to the open pit mines. The affectation of the surface caused by this type of exploitation is also smaller

[3] Jale derives from the Nahuatl word "Xalí" that means "sand" and was adopted in Mexico at the time of the Spanish colonization. In South American countries, the tailings are called "relaves".

because only access routes are constructed such as shafts, ramps, and galleries[4] to reach the mineral and the rest of the exploitation facilities are performed in the underground without altering the surface of the terrain. However, the underground mining activities sometimes produce alterations in the aquifers because of the need to extract ground water to be able to exploit the mineral from greater depths, altering the direction and velocity of the flow, as well as the static level of the water. The mineral production of the underground mines is smaller than in the open pit mines; in Mexico, the underground mines produce from 1000 to 7000 tons of mineral per day.

The wastes generated from the concentration processes are the main source of contamination, as these contain, in most cases, potentially toxic elements (PTE) that under certain condition can be released or incorporated to the environment. The most common release process for these elements is the generation of acid drainage, which occurs by the natural oxidation of minerals containing sulfur when they are exposed to moisture conditions in the presence of oxygen (Sengupa, 1993; Lottermoser, 2007). When acid drainage is produced, the pH reaches levels of 1 to 3, and most PTE are soluble at these pH values, hence, the PTE are dissolved or solubilized and become incorporated as ions in the water current, fostering their dispersion (Environment Australia, 1997).

The acid drainage generation phenomenon occurs not only in mining wastes (tailings and piles) but also in the mining facilities, where sulfide minerals are left exposed; acid drainage can also affect aquifers at the underground constructions.

However, it is important to point out that not all mining wastes generate acidity, since, in some cases, the presence of acid-consuming minerals is larger that that of acid-generating minerals, inducing a neutralization process, which avoids the formation of acid drainage (Sengupa, 1993; Lottermoser, 2007; Environment Australia, 1997).

From the aforementioned, it can be understood that the conditions of each mine are particular to it, and that the environmental impacts produced vary among mines, depending, as mentioned before, on the characteristics of the mineral and the environmental conditions of the site.

In general terms, the environmental impacts produced by the mining activities can be grouped according to their origin and the magnitude of the affectation.

A) Alterations produced in the surface of the terrain

- Removal of vegetation in the mineral exploitation zone (open pit mine), facilities (concentration plant, offices, shops, housing facilities), waste deposits (tailings and piles), access roads.
- Removal of surface soil.
- Alterations of the fauna and its habitat.
- Changes in the topography of the site.
- Alteration of the local hydrographic patterns (runoff patterns of the surface water).

---

[4] A shaft is a vertical construction that serves to access the mine and for mineral extraction. The ramp is an inclined construction (15 to 20°) used to introduce the equipment and services to the mine, as well as to extract the mineral. The galleries are slightly inclined (<2°) a constructions used to access the mine, introduce services, and to extract the mineral.

B) Affectation of the underground and aquifers

- Instability of the terrain due to the mining construction, both underground and on the surface.
- Alteration of the aquifer conditions (direction and speed of the flow and static level) due to water extraction operations.
- Contamination of groundwater due to the dissolution of metals and metalloids from the exposed mineral in the mining facilities or due to infiltration of these elements in solution from the surface stored wastes.
- Contamination of the groundwater due to the generation of acid drainage from sulfide minerals.

C) Affectation of the soil and superficial water bodies

- Contamination of soils and surface water bodies by the dispersion of metals and metalloids contained in the wastes, mainly in tailings.
- By the generation and runoff of acid drainage from the stored wastes.
- Contamination from the materials used in the operation, as is the case of fuels and reagents for processing.
- Contamination from the generated dangerous wastes, mainly oil wastes that originate from the maintenance of the equipment.

D) Other affectations

- Waste water, coming from the operation and housing facilities.
- Waste water from the process, only in very particular cases in which the water is not recycled into the process.
- Solid urban wastes generated by the diverse facilities.
- Noise
- Visual impact
- Vibrations

4. Mining Wastes and Soils Contamination

Tailings and piles represent probably the most important environmental problem of mining activities, since mineral storage and dispersion affect diverse elements of the environment. The risk posed by these wastes lies in their metals and metalloids contents and their potential release.

Tailings are the wastes of greatest concern as potential sources of contamination. During the operation of the concentrating processes, tailings are managed mixed with water forming a pulp, which is sent to the tailings dam, where a water is separated by decantation and sedimentation of tailings; however, the water content within the dam is high and, under certain conditions, can cause fractures in the dam's structure, causing its rupture and release of tailings[5]. Several accidents have been reported around the world due to the rupture of tailings dams, which have brought about severe environmental affectations; one of the best known in the last years was the rupture of the dam in the Aznalcóllar mine in Andalucía, Spain, where 4634 ha of soil were affected, as well as the Agrio and Guadimar rivers (Ericsson and Adamek, 2000). Of the cases reported in Mexico[6] regarding fractures of operating tailings dams, most are related to problems in the management of the water contained in the dam, which brought about overflowing of tailings from the dam, affecting the soil and water in the nearby zones.

Once the tailings dams reach the maximal storage capacity, a closure stage is started, not depositing anymore wastes in it. This starts an evaporation process of the contained water until it dries out. Desiccation of tailings and the size of the particle (less than 150 μm) make easy their dispersion by winds and rainfall. Therefore, in a dam closure process it is necessary to construct facilities to stabilize and cover the material to avoid its dispersion. In diverse mining zones of the country, in which waste deposits were left unsheltered and exposed to weathering for many years, contamination of soil and water occurred due to tailings dispersion; in some cases, due to their metals and metalloids contents, such as arsenic, cadmium, or lead, the contaminated soils have become a risk for the nearby populations (Ongley, 2007; Gutiérrez, *et al.*, 2007; Razo Israel *et al.*, 2004).

The lixiviation piles, although containing also metals and metalloids, induce limited dispersion because the material is constituted by larger sized mineral fragments than that of tailings (from sands to fragments of 4 inches or larger in diameter) that are not easily carried by wind or rainfall. In this case, the environmental risk lies mainly in the potential of the waste to generate acid drainage and release through dissolution of the contained metals and metalloids.

Contamination of soils from the waste deposits represents a risk depending on factors such as the total metals and metalloids concentration, the solubility of these elements according to the mineralogical species, the pH conditions, and the potential for oxidation-reduction in the soil, among others. A series of geochemical processes occur in soils that can limit the mobility of the metals and metalloids once tailings are deposited on the soil. Among these processes, the most important ones are precipitation, co-precipitation, and adsorption (Lottermoser, 2007), which can constitute retention mechanisms that limit metals and metalloids effect on living organisms. These processes, to a large extent, are regulated by the pH and the potential for oxidation-reduction (Eh) (Yong, 2001; Lottermoser, 2007). Due to these processes, sometimes, despite the soil contamination produced by the dispersion of tailings, the risk for living organisms is reduced, because the metals and metalloids contained in the soil are scarcely available, that is, their solubility is low. However, it is important to consider that although the soil-contained metals and metalloids are scarcely available, a change in the pH and Eh conditions can cause their release in soluble form; hence, it is necessary to avoid high soil concentrations of these elements.

Another important aspect to assess the risk posed to the environment by metals and metalloids contaminated soils from mining wastes are the environmental characteristics

[5] The main natural conditions that can cause tailing dam failures are earthquakes, strong rainfalls, and the mechanical characteristics of the soil.

inherent to the site, which vary from one mining zone to another. In mining zones with nearby population settlements, the soil contamination becomes a serious concern for the population itself, for the government authorities, and for those responsible for the generation of wastes. In contrast, in mining zones away from population settlements, the concern is not so strong, and, in general, with time, the natural attenuating mechanisms act incorporating the metals and metalloids to the natural geochemical cycles.

One of the particular characteristics of soil contaminated with mining wastes, particularly from tailings, is the amount of terrain that can be affected mainly by eolian dispersion of tailings from unsheltered deposits. Since eolian dispersion is a cyclic process, accumulation of metals and metalloids in the soil can bring about serious problems in mining zones where no adequate measures have been taken to avoid their dispersion. Assessment of the affectation and application of soil remediation techniques is a difficult and high-cost task because of the large areas of soil that can be affected by tailings.

For this reason, remediation of soils contaminated with metals in mining zones cannot be approached in the same way as in an area where affectation is localized to a limited land lot, and in which the volume of soil to be treated is not high. In the mining zones, first it is necessary to control the dispersion of wastes from the tailings dams or from the lixiviation piles, which per se is a difficult task considering the extension reached by these deposits[7]. The remediation methods to be considered for these cases must be of low cost, due to the large volume of soil that might need treatment. On the other side, the topographical characteristics of the mining sites can complicate and raise the costs of moving the contaminated soils; therefore, *in-situ* treatment would have to be considered.

5. Closure of Mining Waste Deposits

From the environmental point of view, closure of waste deposits is a necessary condition to avoid affectations of soils and water bodies in their vicinity. This closure practice has been applied in Mexico since the start of the last decade, since no precise knowledge was previously available, nor was there any concern, from the mining enterprises, on the effect of these wastes (tailings and lixiviation piles) on the environment, even after ending the exploitation of the mine.

Closure of a wastes deposit implies fulfilling two basic conditions: a) to avoid dispersion of the waste by wind or water and b) to warrant the stability of the deposit.

Several techniques have been developed worldwide to control dispersion of wastes and to control the release of metals and metalloids through the generation of acid drainage; most of these techniques are aimed at covering the wastes and avoid the oxidation process by eliminating the factors that promote it (mineral with sulfides, water, and oxygen).

To confine the mining wastes, the following types of covers have been used (Environment Australia, 1997; Lottermoser, 2007; US EPA, 1994; Sengupa, 1993):

---

[6] These cases were reported to the National Water Agency (CNA, for its initials in Spanish) and were published in the web page of the Federal Commission for Regulations Improvement (COFEMER, for its initials in Spanish; see references).

[7] In Mexico, the tailing dams and the piles occupy up to several hundreds of hectares, making it difficult to control 100% the dispersion of wastes.

- Geotextiles (synthetic membranes)
- Water (to saturate the waste)
- Clayish matter
- Soil with organic matter
- Neutralizing materials (such as limestone rock)
- A combination of the former.

The use of any of these depends on diverse factors, standing out, among them, the inherent characteristics of the waste (its risk), availability of the cover material (in stock in nearby sites), and, above all, the costs. In developed countries, like the USA, Canada, and Australia, covers of synthetic material (geomembranes) have been used or saturation with water (this means flooding the waste deposits) to avoid dispersion. In most cases, these measures are highly expensive due to the large volume of stored wastes and the large size of the occupied terrain. In Mexico, for example, covering with geomembranes is not used because of its high cost, and water flooding is barely available because many of the mining centers are located in arid zones. Covering with clayish material or neutralizing agents depends on the availability of these materials in zones nearby the deposits, because their transport from far away sites increases the costs considerably. In Mexico, the use of sterile material, coming from the same mineral extraction process, and of soil with organic matter, removed previously from other operation sites, has been chosen. This practice has the advantage of contributing to the formation of a soil layer on top of the wastes that in time becomes enriched with nutrients, organic matter, and microorganisms, allowing it to sustain the growth of plants.

For the natural materials cover to be effective it is necessary to perform reconfiguration works in the wastes deposit to handle rain water runoffs and to reduce to a maximum the erosion of the fertile soil layer. These reconfiguration works deal with reducing the inclination of the slopes, strengthening the foot of the slopes, construction of berms at the foot of the slopes, and providing an inclination towards the inside so that the water will not runoff on the sides of the slope, as well as constructing perimetric channels to lead the pluvial water, among others (Parra and Perez, 2003; Muhech, 2001).

In this way, revegetation of the piles and tailings dams is a requisite to avoid and control dispersion of wastes. Revegetation is an activity that must be carefully planned, considering that the best suited vegetal species are those endemic of the region; in this way, the possibilities for their adaptation to the climatic conditions, soils, and to the waste itself are much better, allowing for their adequate development; besides, the use of native species permits to reestablish the natural biological cycles and, hence, the waste deposits can be incorporated to the current conditions of the environment.

Revegetation of waste deposits contributes markedly to their stability, since plants avoid dispersion of tailings by the wind, serve as retention medium, and contribute to reduce their being carried away by rainfall runoff. Another important aspect of revegetation is the capacity of some plants to adapt to the presence of certain contaminants, organic or inorganic (US EPA, 2000; Australian Government, 2006). The presence of metals and metalloids in waste deposits can limit the growth and development of plants, since, in some cases, these elements can be toxic to some species; however, in other cases, these elements can be absorbed and accumulated at high concentrations by the plants (Chaney *et al.*, 2000; McGrath *et al.*, 2000).

Although this accumulation mechanism is favorable for soil phytoremediation methods, it is not so for the chemical stabilization of mining waste deposits. These deposits, with time, attract herbivore animals, for example grazing animals, which, by ingesting these plants can become affected and, in this way, the metals and metalloids contained in the plants would incorporate into the trophic chain. However, no studies have been performed yet to assess the toxicological risk posed by metals and metalloids accumulating plants growing on mining waste deposits for herbivore animals that feed on these sites. For this reason, it is important to assess the plants that can be used in the revegetation of mining waste deposits and search for those that do not accumulate metals and metalloids in their tissues, but rather bind them in their roots and immobilize them.

Aside from the physical stability attained with the cover, the reconfiguration of the deposit and revegetation, it is important to warrant the chemical stability of the wastes regarding the release of solubilized metals and metalloids, which induces acid drainage. In those cases in which the processed mineral is mainly composed of metal sulfides, the wastes can become potential generators of acid, unless the presence of acid-consuming minerals, such as calcite, aluminum silicate, oxihydroxides, and clayish minerals, is higher and can therefore neutralize the generated acidity.

In these cases it will be necessary to apply complementary measures to control the acidity once the oxidation of sulfides has started. To this regard, diverse technologies have also been developed worldwide to control the generation of acid drainage or to treat it, once generated. Among these are the aerobic and anaerobic wetlands, anoxic limestone drainages, successive alkalinity production systems, open limestone channels, and permeable reactive barriers (López, 2002; Filipek *et al.*, 2003).

In Mexico, no reports exist on the use of any of the mentioned methods to control the generation of acid drainage; this opens the door to a research field of great interest for the country.

From the aforementioned, it can be observed that the closure of a mining wastes deposit, either a tailings dam or a lixiviation pile, bears a series of measures that complement each other to reduce to a maximum the dispersion of wastes and the release of metals and metalloids contained in them. It should be pointed out that the cost of these measures is high; therefore, the closure of a wastes deposit must already be carefully planed during the operational stage of the mine.

6. Alternatives for Soils Remediation

Remediation of metals and metalloids contaminated soils from mining wastes is a task presenting, generally, a series of limitations for the application of conventional soil treatment methods[8], given that occasionally the zones are poorly accessible, or due to the large affected surface, or by the high cost of some of the technologies. In the particular case of mining zones, phytoremediation methods might be efficacious given the conditions in which the soil affectation occurs. For example, it is common to find low concentrations of metals and metalloids in the areas neighboring a wastes deposit, or the contaminants are restricted to the surface layer of the soil (0 to 20 cm of depth), or even sites exist that do not pose a direct risk for the nearby populations. Phytoremediation is based on the use of plants capable of removing or transforming the soil contaminants, through diverse processes, mainly absorption, microbial degradation, or stabilization by phase changes (Reddi and Inyang, 2000).

The mechanisms developed by plants are diverse, in some cases the contaminant is absorbed by the root and translocated to the stem, leaves, and flowers, accumulating in each part of the plant (*phytoextraction*). Another studied mechanism is the adsorption of contaminants by the root of the plants, thereby limiting their mobilization, but without letting the contaminant reach other parts of the plants (*phytostabilization*); in other cases, the microorganisms present in the root of the plant degrade the contaminants, turning them into substances less damaging to the environment (*rhizodegradation*); the enzymes produced by the plants can also unfold some contaminants through metabolic processes (*phytodegradation*); transpiration of plants is also a mechanism by which some volatile contaminants are taken up from the soil and released into the atmosphere (*phytovolatilization*).

In the mining zones, it is common to observe some species that grow on sites potentially affected by the mining wastes, and there are even species able to strive on the waste deposits; thus, it can be inferred that these species developed some mechanism to adapt to the presence of metals and metalloids in the soil. Diverse studies performed around the world show that some plant species are able to accumulate variable amounts of metals and metalloids in mining zones (Chang *et al.*, 2005; Visoottivisetha *et al.*, 2002; Li *et al.*, 2007; Carrillo and González, 2006; Clemente, 2006). However, not all plant species can accumulate large PTE concentrations, as do hyperaccumulating species, some are only tolerant to the presence of contaminants, and some are not even able to strive because the contaminant acts as an inhibitor of their development.

To use plants for soil phytoremediation in mining zones, it is important to identify species inherent to the region (native plants) able to accumulate the largest amount of contaminants or capable of fixing them to the soil. Once these species have been identified it is necessary to reproduce them in nurseries to transplant them afterwards into the zones identified as highly metals and metalloids contaminated soils. It is important to limit land lots to isolate them to hinder grazing by herbivore species that would destroy plants growing on them. In several countries, phytoremediation has been applied successfully for the extraction and stabilization of metals, such as copper, zinc, cadmium, nickel, cobalt, manganese, and lead, in contaminated soils (Baker *et al.*, 2000; Chaney *et al.*, 2000).

In Mexico, attempts have been made with phytoremediation, basically using common plants, such as sunflower, lettuce, and zucchini, among others (Beltran *et al.*, 2006). In mining zones, plants that grow near the waste deposits or on the deposits themselves have been studied, but no soil phytoremediation applications have been reported yet with native plants from the mining sites, which orients toward a very interesting research line with a large field of application.

Phytoremediation has shown good results, particularly, in metals-contaminated soils from the mining waste deposits. Diverse studies show the capacity of plants to accumulate metals present in soils from mining zones (Carrillo, 2005; Carrillo and González, 2006; Li *et al.*, 2007; Chang *et al.*, 2005; Clemente *et al.*, 2006).

---

[8] Methods such as soil washings, neutralization, vitrification, incineration, or confinement.

In Mexico, plants that reproduce in mining zones have been studied (Díaz *et al.*, 2005; González *et al.*, 2005), finding that they are able to retain certain metals such as cadmium and zinc. It has also been found that some plants growing near tailings dams are able to accumulate significant amounts of arsenic and zinc (Puga *et al.*, 2006). Other investigations have focused on metal removal with vegetal species, although not in soils from mining zones (Beltrán *et al.*, 2006; Rodríguez Ortiz *et al.*, 2006).

In the particular case of mining zones, once enough time has elapsed for the plants to accumulate a considerable amount of metals and metalloids, a composting method can be developed with them to degrade the organic part, and use it to fertilize the soil used in the cover of the wastes deposit. In this way, the PTE accumulated in the plants would be returned to the wastes deposit from which they originally dispersed. Another alternative would be to process the material produced by composting to recover the metals contained in the compost, although this is not economically attractive due to the low contents to be extracted from this material.

7. Case Study

In the municipality of Cadereyta, state of Querétaro, Mexico, there is a mine that exploits a polymetallic ore deposit, mainly lead, copper, and zinc. The wastes generated in this mine contain significant amounts of these elements, aside from other, such a arsenic, cadmium, and iron. During the operational time of this mine, wastes have been accumulated in five dams. These storage deposits represent a source of contamination because some parts of them are unsheltered and exposed to the meteorological conditions of the site (Figure 1).

In the oldest dams, vegetation has been induced to avoid erosion of the waste, but vegetation inherent to the region has developed, which has adapted to the presence of the metals contained in the waste.

In this mine, a study was started on the vegetal species that have been able to develop on the waste deposits to assess their capacity to stabilize the metals contained in the waste and use them for revegetation of the currently active dams, as well as to develop a remediation method that will permit to clean up the soils previously affected by the dispersion of the wastes (Figure 2).

The methodology applied in this study started by identifying the predominant species growing on the tailings dams. Then, representative samples of each species from different sites were collected. On one side, for their taxonomic classification and, on the other, to determine the concentration of metals and metalloids in the different parts of the plant (root, stem, leaves). Likewise, a sample of soil from the site of each plant was taken and, in some cases, from the waste as well. According to the composition of the mineral extracted in this mine, the metal elements identified in the samples of plants, soils, and wastes were arsenic, lead, cadmium, copper, and zinc.

Ten species of plants were identified as predominating in the waste deposits and are presented on Table 1.

Some of the identified species belong to families that have already been reported in other works as being able to develop on mining waste deposits, as for example, *Leguminosae, Solanaceae, Casuarinaceae* (Carrillo and González, 2006). However, other species are inherent to this region and have reached the capacity to grow on these sites; their relevance lies in that they might reproduce with more ease under the climatic conditions of the site.

Figure 1. Tailings dam in La Negra mine.

Figure 2. Plants that grow in tailing dam

Table 1. Species of plants in tailings dam

Family	Genus	Specie	Author
Solanaceae	*Nicotina*	*glauca*	Graham
Asteraceae	*Flaveria*	*pubescens*	Rydb.
Anacardiaceae	*Schinus*	*molle*	L.
Casuarinaceae	*Casuarina*	*Sp.*	
Bignoniaceae	*Tecoma*	*stans*	(L.) Kunth
Leguminosae	*Prosopis*	*Sp.*	
Poaceae	*Cenchrus*	*ciliaris*	L.
Scrophulariaceae	*Maurandia*	*antirrhiniflora*	Humb.& Bonpl. ex Willd
Euphorbiaceae	*Ricinus*	*cummunis*	L.
Cactaceae	*Opuntia*	*lasiacantha*	Pfeiffer
Cactaceae	*Opuntia*	*tomentosa*	Salm-Dyck

Table 2 presents the species with the highest concentration in their roots (R), stems (S), branches (B), leaves (L), and flowers (F).

It was found that the species that accumulate significant amounts of arsenic were *Nicotina glauca* with up to 42.91 mg/kg of As in branches and 46.13 mg/kg in leaves (JS 009). Other important species were *Flaveria pubescens*, which, although small-sized, is able to accumulate up to 15 mg/kg of As (JS 019) and *Tecoma stans* that accumulates up to 22.48 mg/kg of As in branches and 30.13 mg/kg in leaves (JS 016). The species *Cenchrus ciliaris*, a grass, is also interesting, as it retains, in the root, up to 8.13 mg/kg (JS 010); likewise *Opuntia lasiacantha* accumulates in the root up to 12.54 mg/kg of As (JS 024).

The same species, *N. glauca*, *F. pubescens*, *T. stans*, are important accumulators of cadmium. *N. glauca* accumulates in the branch up to 42.48 mg/kg of Cd (JS 009) and in leaves up to 51.05 mg/kg (JS 009); *F. pubescens* accumulates up to 11.58 mg/kg in leaves (JS 019) and *T. stans* accumulates up to 25.37 mg/kg in branches and up to 32.80 mg/kg in leaves (JS 016).

Regarding Pb, which is a poorly soluble element in dams (it is found as PbS, a very stable mineral), its concentrations in plants was low.

Finally, for Zn, the same species, *N. glauca*, *F. pubescens*, *T. stans* have the capacity to concentrate this element. Sample JS 009 of *N. glauca* accumulated 905.64 mg/kg in the branch. Sample JS 019 of *F. pubescens* had a concentration of 579.84 mg/kg of Zn in the leaf. Sample JS 016 of *T. stans* had 927.03 mg/kg of Zn in the leaf. For zinc, the species *C. ciliaris* was also interesting, sample JS 010 had a concentration of 280.89 mg/kg in leaves; likewise *O. lasiacantha* (JS 024) accumulated zinc in the root (413.17 mg/kg) but does not translocate it to the rest of the plant.

Another interesting aspect of the results is the fact that large size arboreal species, such as the "pirú" (*Schinus molle*), the "mezquite" (*Prosopis Sp.*), and the pine (*Casuarina Sp.*), the latter is an induced species, do not store metallic elements in significant amounts; therefore, their presence in the tailings dams contributes rather to avoid dispersion of tailings and to provide stability to the deposit than to fix the metallic elements or to decrease their release.

Table 2. Metals and metalloids concentration in plants.

Sample	Concentration in mg/kg					Specie
	As	Cu	Cd	Pb	Zn	
JS 009-R	8.06	10.48	8.48	58.47	259.43	*Nicotina glauca*
JS 009-S	8.49	ND	10.28	14.68	224.64	
JS 009-B	42.91	14.39	47.48	44.48	905.64	
JS 009-L	46.13	51.45	51.05	44.47	622.13	
JS 009-F	18.00	41.82	17.65	63.78	332.93	
JS 010-R	8.13	18.61	9.21	55.24	317.25	*Cenchrus ciliaris*
JS 010-L	3.71	11.40	4.00	47.38	280.89	
JS 016-R	3.63	9.89	5.20	59.86	138.92	*Tecoma stans*
JS 016-S	2.88	0.90	4.99	30.96	171.79	
JS 016-B	22.48	14.29	25.37	58.44	564.44	
JS 016-L	30.13	52.73	32.80	73.27	927.03	
JS 019-R	4.54	25.36	3.99	51.52	219.65	*Flaveria pubescens*
JS 019-B	2.32	12.97	3.49	67.84	103.75	
JS 019-L	15.06	88.32	11.58	133.73	579.84	
JS 019-F	4.29	37.22	5.79	70.35	586.71	
JS 024-R	12.54	13.77	11.68	36.13	413.17	*Opuntia lasiacantha*
JS 024-L	0.33	ND	0.30	23.96	67.98	

One of the most relevant aspects to consider in these species is the specific site at which they were collected, that is, whether they were taken from the inside of the dam or from the slope, because in the slope the erosion process is stronger and this limits the growth of plants due to a continuous loss of substrate. In the impoundment, the runoff is smaller and, in general, the water concentrates there.

Regarding the physical stability of the deposit, it is important to combine on the slope the larger species of deeper-extending roots with shrub-like species. In this case, the function of the larger-sized species is to fix the fertile soil substrate to the waste, whereas the main function of the shrub-like species is to reduce the erosion processes.

Another important aspect is the potential generation of acidity inherent to this waste. In one of the deposits, the waste in the slope was already in an advanced oxidation process and the pH at this site was acid; however, samples of grass (*Cenchrus ciliaris*) were taken and, interestingly, this species is capable of adapting also to the acid conditions of the substrate, furnishing it with an additional advantage.

Determination of the metals concentration in the plants will allow identifying their capacity to reduce the mobility of these metal species, as well as to perform a preliminary assessment of those plants that can be used to remediate soils contaminated with wastes in the zone near the dams.

REFERENCES

[1] AIMMGM. 1993. Manual de presas y depósitos de jales. Asociación de Ingenieros de Minas, Metalurgistas y Geólogos de México, A.C. México, D.F.80 pp.

[2] Armienta M.A., Talavera O., Morton O., Barrera M. 2003. Geochemistry of Metals from Mine Tailings in Taxco, Mexico. Bull. *Environmental Contamination and Toxicology.* 71:387-393.

[3] Australian Government. 2006. Mine rehabilitation. Department of Industry, Tourism and Resources. *Australian Government.* ISBN 0 642 72481 4.

[4] Baker Alan J.M., McGrath S.P., Reeves Roger D., Smith J.A.C. 2000. Metal hyperaccumulator plants: a review of the ecology and physiology of a biological resource for phytoremediation of metal-polluted soils. In: Phytoremediation of contaminated soils and water. Norman Ferry, Bañuelos Gary (editors). Lewis Publishers Inc. USA. 389 pp.

[5] Beltrán Margarita, Vaca Mabel, López Raymundo, Alcalá Juan. 2006. Phytoextraction of heavy metals and transport and distribution in plant tissue. Environmental Biotechnology and Engineering. Proceedings of the Second Internacional Meeting on Environmental Biotechnology and Engineering. 26-29 September, 2006. Mexico City.

[6] Carrillo González Rogelio. 2005. Niveles de contaminación de los suelos y las plantas. En: El sistema planta-microorganismo-suelo en áreas contaminadas con residuos mineros. González Chávez M. C., Pérez Moreno J. y Carrillo González R. (editores). Colegio de Posgraduados. Montecillo, Estado de México. 161 pp.

[7] Carrillo González R., González Chávez C. 2006. Metal accumulation in wild plants surrounding mining wastes. *Environmental Pollution*, 144 (2006) 84-92.

[8] Castro Larragoitia J., Kramer U., Puchelt H. 1997. 200 years of mining activities at La Paz/San Luis Potosí/Mexico – Consecuentes for environment and geochemical exploration. *Journal of Geochemical Exploration*. 58 81-91.

[9] Clemente Rafael, Almela Concepción, Bernal M. Pilar. 2006. A remediation strategy based on active phytoremediation followed by natural attenuation in a soil contaminated by pyrite waste. *Environmental Pollution* 143 397-406.

[10] CNA. 2000. Manual de presas de jales. Subdirección General Técnica, Gerencia de Ingeniería Básica y Normas Técnicas. Comisión Nacional del Agua. Mexico, D.F.

[11] COFEMER. Comisión Federal de Mejora Regulatoria. Portal de Manifestaciones de Impacto Regulatorio.

[12] www.cofemermir.gob.mx/uploadtests/2469.66.59.2.Anexo%20B.doc

[13] Chaney Rufus L., Li Yin-Ming, Brown Sally L., Homer Faye A., Malik Minnie, Angle Scott J., Baker Alan J.M., Reeves Roger D., Chin Mel. 2000. Improving metal hyperaccumulator wild plants to develop commercial phytoextraction systems: approaches and progress. In: Phytoremediation of contaminated soils and water. Norman Ferry, Bañuelos Gary (editors). Lewis Publishers Inc. USA. 389 pp.

[14] Chang Peichung, Kim Ju-Yong, Kim Kyoung-Woong. 2005. Concentrations of arsenic and heavy metals in vegetation at two abandoned mine tailings in South Korea. *Environmental Geochemistry and Health* (2005) 27:109-119.

[15] Díaz Garduño Lucina M., Díaz Osornio Agustina, Carrillo González Rogelio, González Chávez María del Carmen. 2005. Plantas que se desarrollan en áreas contaminadas con

residuos mineros. En: El sistema planta-microorganismo-suelo en áreas contaminadas con residuos mineros. González Chávez M. C., Pérez Moreno J. y Carrillo González R. (editores). Colegio de Posgraduados. Montecillo, Estado de México. 161 pp.

[16] Environment Australia. 1995. Mailing Containment. *Best Practice Environmental Management in Mining*. ISBN 0 642 19423 8.

[17] http://www.natural-resources.org/minerals/cd/docs/ea/booklets/tailings/tails1.pdf

[18] Environment Australia. 1997. Managing sulphidic mine wastes and acid drainage. *Best Practice Environmental Management in Mining*. ISBN 0 642 19449 1.

[19] http://www.natural-resources.org/minerals/cd/ea_sulph.htm.

[20] Ericsson N. and Adamek P. 2000. The tailings pond failure at the Aznalcóllar mine, Spain. Sixth International Symposium in Environmental Issues and Waste Management in Energy and Mineral Production. Calgary, Alberta, Canada, 30 May – 2 June 2000.

[21] Filipek L.H., Hatton G., Gusek J., Tsukamoto T. 2003. Passive treatment of acid rock drainage (ARD): state of the practice. In: Tailings and Mine Waste '03. A. A. Balkema Publishers. Netherlands. 339-343

[22] García Meza J.V., Ramos E., Carrillo Chávez A., Durán de Bazua C. 2004. Mineralogical and chemical characterization of historical mine tailings from the Valenciana mine, Guanajuato, Mexico: Environmental implications. Bull. Environmental Contamination and Toxicology. (2004) 72:170-177.

[23] González Chávez María del Carmen, Carrillo González Rogelio, Pérez Moreno Jesús. 2005. Perspectivas en la recuperación de áreas contaminadas con residuos mineros utilizando plantas y microorganismos simbióticos. En: El sistema planta-microorganismo-suelo en áreas contaminadas con residuos mineros. González Chávez M. C., Pérez Moreno J. y Carrillo González R. (editores). Colegio de Posgraduados. Montecillo, Estado de México. 161 pp.

[24] Li M.S., Luo Y.P., Su Z.Y. 2007. Heavy metal concentrations in soils and plant accumulation in a restored manganeso mineland in Guangxi, South China. *Environmental Pollution* 147 (2007) 168-175.

[25] López Pamo E., Aduvire O., Barettino D. 2002. Tratamientos pasivos de drenajes ácidos de mina: estado actual y perspectivas de futuro. Boletín Geológico y Minero, 113 (1): 3-21

[26] Lottermoser Bernd. 2007. Mine Wastes. Characterization, treatment and environmental impacts. Second Edition. Springer. U.S.A. 304 pp.

[27] McGrath S.P., Dunham S.J., Correl R.L. 2000. Potential for phytoextraction of zinc and cadmium from soils using hyperaccumulator plants. In: Phytoremediation of contaminated soils and water. Norman Ferry, Bañuelos Gary (editors). Lewis Publishers Inc. USA. 389 pp.

[28] Mendoza Amézquita E., Armienta Hernández M.A., Ayora Carlos, Soler Albert, Ramos Ramírez E. 2006. Potencial lixiviación de elementos traza en jales de las minas La Asunción y Las Torres, en el Distrito Minero de Guanajuato, México. Revista Mexicana de Ciencias Geológicas. Vol. 23, Núm. 1, 2006, 75-83.

[29] Muhech Dip V. 2001. Acciones sugeridas para el abandono ordenado de almacenamiento de residuos mineros. Memorias del Seminario "Almacenamiento de Residuos Mineros". Sociedad Mexicana de Mecánica de Suelos. México, D.F. 131-137.

[30] Ongley Lois K., Sherman Leslie, Armienta Aurora, Concilio Amy, Ferguson Salinas Carrie. 2007. Arsenic in the soils of Zimapán, Mexico. Environmental Pollution 145 (2007) 793-799.

[31] Parra H.I. and Perez M. A., 2003. Abandonment of tailings deposit Casapalca. In: Tailings and Mine Waste '03. A. A. Balkema Publishers. Netherlands. 339-343.

[32] Puga Soraya, Sosa Manuel, De la Mora Antonio, Pinedo Carmelo, Jiménez Jorge. 2006. Concentraciones de As y Zn en vegetación nativa cercana a una presa de jales. Revista Internacional de Contaminación Ambiental 22 (2) 75-82, 2006.

[33] Razo Israel, Carrizales Leticia, Castro Javier, Díaz-Barriga Fernando, Monroy Marcos. 2004. Arsenic and heavy metal pollution of soil, water and sediments in a semi-arid climate mining area in Mexico. *Water, air and soil pollution* 152: 129-152 2004.

[34] Reddi Lakshmi N., Inyang Hilary I. 2000. Geoenvironmental Engineering, principles and applications. Marcel Dekker Inc. USA. 494 pp.

[35] Ripley Earle, Redmann Robert, Crowder Adele. 1996. Environmental Effects of Mining. St. Lucie Press. USA. 356 pp.

[36] Rodríguez Ortiz Juan Carlos, Rodríguez Fuentes Humberto, De Lira Reyes Gerardo, Martínez de la Cerda Jesús, Lara Mireles José Luís. 2006. Capacidad de seis especies vegetales para acumular plomo en suelos contaminados. Revista Fitotecnia Mexicana. Vol. 29 (3): 239-245, 2006.

[37] Romero Francisco M., Armienta M. A., Villaseñor G., and González J.L.. 2006. Mineralogical constraints on mobility of arsenic in tailings from Zimapan, Hidalgo, Mexico. *Int. J. Environment and Pollution*, Vol. 26 (1/2/3): 23 – 40.

[38] Romero Francisco M., Armienta M.A., González Hernández G. 2007. The solid-phase control on the mobility of potentially toxic elements in an abandoned lead/zinc mine tailings impoundment, Taxco, Mexico. *Applied Geochemistry*, Vol. 22. 109-127.

[39] Gutiérrez R. Margarita, Romero Francisco, González H. G. 2007. Suelos y sedimentos afectados por la dispersión de jales inactivos de sulfuros metálicos en la zona minera de Santa Bárbara, Chihuahua, México. Revista Mexicana de Ciencias Geológicas. Vol. 24 No. 2, 170-184.

[40] Sengupa Mritunjoy. 1993. Environmental impacts of mining. Monitoring, restoration and control. Lewis Publishers. USA. 494 pp.

[41] US EPA. 1994. Design and evaluation of tailings dams. Office of Solid Waste. U.S. *Environmental Protection Agency*. EPA/530/R-94/038.

[42] US EPA. 2000. Introduction to phytoremediation. Office of Research and Development. *U.S. Environmental Protection Agency*. EPA/600/R-99/107.

[43] Visoottivisetha P., Francesconib K., Sridokchana W. 2002. The potential of Thai indigenous plant species for the phytoremediation of arsenic contaminated land. *Environmental Pollution*. 118 (2202). 453-461.

[44] Yong Raymond N. 2001. Geoenvironmental engineering contaminated soils, pollutant fate and migration. CRC Press. U.S.A. 307 pp.

In: Remediation of Soils and Aquifers
Editors: Luis G. Torres and Erick R. Bandala

ISBN 978-1-60741-372-1

Chapter 13

SOIL WASHING FOR REMEDIATION OF SOILS CONTAMINATED WITH PETROLEUM HYDROCARBONS AND/OR METALS

Luis G. Torres * ***and Erick R. Bandala***
UPIBI-IPN and UDLA-Puebla

ABSTRACT

Two industries which have caused constant contamination problems in Mexico are the petrochemical and the mining industries. In both cases, industrial activity has lead to the contamination of a number of sites in the national territory (more than 76 sites are contaminated with metals and more than 103 with petroleum derivatives) (SEMARNAT 2000).Surfactant enhanced soil washing (SESW) has become a very important option for the remediation of soils contaminated with a variety of compounds and elements, i.e., petroleum hydrocarbons, pesticides, dense nonaqueus-phase liquids, metals, and others. Soil washing enhanced by surfactants is a feasible remediation technique when there is no chance for biodegradation-based methodologies. In the case of soils contaminated with metals, there are no degradation process, since metals cannot be destroyed, but isolated, removed or stabilized, among other physical process.When time and/or space are scarce, soil washing is an excellent cost/benefit option. This chapter reviews the state of the art for surfactant enhanced soil washing developments. Aspects such as the selection of the best surfactant and doses and the use of natural surfactants, alone or in combination with synthetic ones, are discussed. The specific issues when washing soils contaminated with metals are also discussed. Finally, aspects related with the design of an industrial washing process and estimation of costs are reviewed and discussed.

ORIGIN OF THE PROBLEM

Two industries which have caused constant contamination problems in Mexico are the petrochemical and the mining industries. In both cases, industrial activity has lead to the

* LTorresB@ipn.mx, LTorresB@iingen.unam.mx

contamination of a number of sites in the national territory (more than 76 sites are contaminated with metals and more than 103 with petroleum derivatives) (SEMARNAT 2000).

Surfactant-enhanced soil washing has become a very important option for the remediation of soils contaminated with a variety of compounds and elements, *i.e.,* petroleum hydrocarbons, pesticides, dense nonaqueus-phase liquids, metals and others. Soil washing enhanced by surfactants is a feasible remediation technique when there is no chance for biodegradation-based methodologies. When time and/or space are scarce, soil washing is an excellent cost/benefit option.

This chapter reviews the state-of-the-art of surfactant enhanced soil washing developments for soils contaminated with petroleum hydrocarbons and metals in some extent. However, information presented and discussed here can be applied to other cases, such as soils contaminated with pesticides, pharmaceutical products, PCBs, halogenated organics, etc.

Basically, soil washing is defined as a physicochemical separation technology in which excavated soil is pretreated to remove large objects and soil clods and then washed with fluids to remove contaminants. To be effective, soil washing must either transfer the contaminants to the wash fluids or concentrate the contaminants in a fraction of the original volume, using size separation techniques (USEPA, 1991).

Some Issues Regarding Surfactants

Firstly, surfactants are molecules with the characteristic of presenting a polar head and a non-polar tail, which imparts both hydrophobic and hydrophilic characters to the molecule. For the SESW, some basic aspects regarding surfactants are of vital importance. At least a general knowledge of the chemical nature, molecular weight, hydrophobic-hydrophilic balance, and the critical micelle concentration for most common surfactants is necessary. Chemical nature is quite important, because a surfactant is ionic or nonionic, cationic or anionic, and the specific chemical structure affects the soil washing performance. Table 1 lists some surfactants including this information.

Specifically, two parameters are of crucial importance. The hydrophilic-lipophylic balance, HLB, is a measure of the dual behavior of any surfactant: a portion of the molecule is water soluble, whereas the rest of the molecule is non-polar (i.e., it shows more affinity for the oily phase). Products with low HLB are prone to form water in oil emulsions (they are more lipophylic), whereas surfactants with high HLB values tend to form oil in water emulsions (they are more hydrophilic). The second important parameter is the critical micelle concentration (CMC) which, by definition, is the surfactant concentration at which surfactant monomers start to micellize. This means that every surfactant, due to its intrinsic properties, will show a different CMC value.

Finally, it is very interesting to mention that the CMC of ionic surfactants can be modified due to the presence of salts (especially monovalent salts, i.e., Na^+, K^+). This effect has been widely studied. For this review, it is only necessary to say that the CMC value for a given surfactant decreases when salt concentration is enhanced, up to a limit value. No modification will be observed even when more and more salt is added, over this limit value, to the system. This fact is shown in figure 1, where the CMC value for SDS was measured in presence of 0.0, 0.5, 1, and 2% NaCl (Torres et al., 2005). A very enlightening work for those interested in CMC features is that published by Huibers et al. (1996), in which prediction of CMC is achieved through a quantitative structure-property relationship approach.

Table 1. Some surfactants and their main properties (ionic nature, chemical name, mol weight, HLB and CMC values).

.Surfactant	Ionic nature	Chemical name	Mol weight (g/g mol)	HLB	CMC (mg/L)
SDS	Anionic	Sodium dodecyl sulphonate	288.4	40	400
Texapon 40	Anionic	Sodium lauryl ether sulphate	442	NR	1,458
Maranil Lab	Anionic	Sodium dodecyl-bencen sulphonate	348	NR	1,392
Emulgin 600	Non-ionic	Nonyl phenol (Poe 6)	483	11.0	45.06
Emulgin 1000	Non-ionic	Nonyl phenol (Poe 10)	NR	13.5	NR
Brij 35	Non-ionic	Lauric alcohol ether (Poe 23)	1,206	16.7	NR
Brij 58	Non-ionic	Stearilic alcohol ether (Poe 20)	1,122	5.6	NR
Brij 72	Non-ionic	Stearilic alcohol (Poe 2)	358	4.9	NR
Span 20	Non-ionic	Sorbitan monolaurate	214	8.6	NR
Tween 20	Non-ionic	Sorbitan monolaurate (Poe 20)	1,226	16.7	60.74
Tween 60	Non-ionic	Sorbitan monostearate (Poe 20)	1,310	14.9	NR
Tween 80	Non-ionic	Sorbitan monoleate (Poe 20)	1,308	15.0	65.4

*Adapted from Torres et al. (2004).

Tests Focused in the Selection of Proper Surfactants and Doses

These assessments are tests carried out to determine basic aspects at the laboratory level. Many works deal with the selection of the right surfactant or mixture of surfactants and the optimal doses for treating a given soil contaminated with certain hydrocarbon fractions or metals present in the soil (Butler and Hayes, 1998; Deshpande et al., 1999; Deshpande et al., 2000; Kuei-Jyum and Young, 2003; Torres et al., 2003; Lopez et al., 2004; Torres et al., 2007a, b and c; Ahn et al., 2008; Zhou et al., 2005; Vreysen and Maes, 2005). Other works deal with aspects covering the presence of fine and coarse fractions (Kuei-Jyum and Young, 2003), the use of salts to enhance the washing process, especially when using ionic surfactants (Lopez et al., 2003; Torres et al., 2005), the relevance of the properties of the contaminant over the soil washing procedure (Torres et al., 2007c) and other topics (See figure 2).

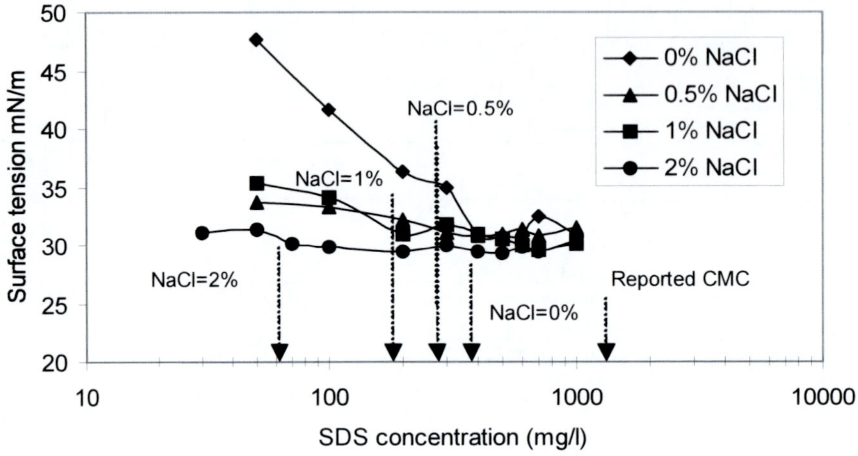

Figure 1. CMC for SDS in presence and absence of NaCl (Adapted from Torres et al., 2005).

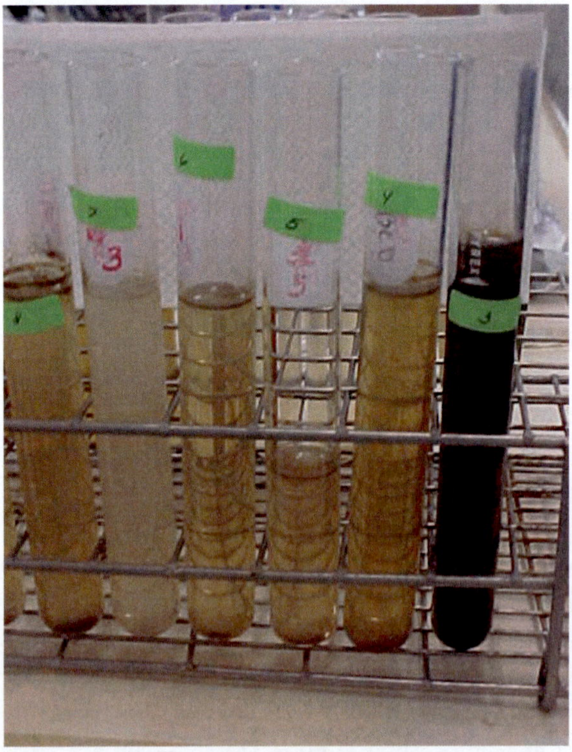

Figure 2. Anionic and non-ionic surfactant solutions after washing a crude-contaminated soil (Photograph: L. Torres).

Some works explore the use of natural surfactants or mixtures of natural-synthetic surfactants to attain the desired efficacy in the soil washing process, with the additional advantage of producing more biodegradable wastewater than those obtained when using only synthetic surfactants (Berselli et al., 2004; Torres et al, 2007a; Urum and Pekdemir, 2004; Urum et al., 2005). The use of guar gum, locust bean gum (or carubina), cyclodextrine, humic substances, rhamnolipids, soy lethicine, aescine, saponin, and tannin -among others- has been reported as a good alternative to synthetic surfactants, such as SDS, Tween 80, Surfacpol 600, or several others, for soil washing.

Torres et al. (2003) reported three surfactants with diesel-removal efficiencies in petroleum-contaminated soils. These surfactants were ethoxylated sorbitan monoleate, ethoxylated nonylphenol, and sodium lauryl ether sulfate. After the measurement of the CMC values for the three surfactants, which were 65.4, 45.06, and 414 mg/l, respectively, their capabilities of removing TPH-diesel fraction from a Mexican soil were measured. The three surfactants resulted very useful for SESW of that kind of contaminated soil, giving high removals, as those shown in Table 2. It is remarkable that using sub-CMC concentrations (0.5x CMC), TPH-diesel removal efficiencies higher than 46, 41, and 52%, respectively, were obtained. For the supra–CMC concentration, using 50, 100, and 300 times the CMC concentration, TPH removal efficiencies up to 74, 83, and 88%, respectively, were achieved.

More recently, Torres et al. (2007) published a work dealing with the following question: Do characteristics of crude oil in contaminated soils affect its removal by washing? The definitive answer was: yes! In this study, the removal efficiencies found when washing soils contaminated with three different crude types (at approximately the same concentration) with four different solutions, i.e., water, SDS solution, SDS+NaCl solution, and SDS+Na-metasilicate solution were quite different. Before the SESW assessments, the soils were characterized in terms of TPH, BTEX, PAHs, TPH_{diesel}, $TPH_{gasoline}$, total metals, Ca+Mg, and API density. TPH removal efficiencies for each washing solution and each contaminated soil were correlated; the final result was that the contaminated soil properties that correlated well were, in descending order: $TPH_{gasoilne}$ = crude oil density = BTEX > TPH_{diesel} = Ca+Mg > TPH. Total metal concentration and PAH never correlated with the amount of contaminant removed by washing.

Remediation of Soils Contaminated with Metals

The use of soil washing for soils contaminated with metals has been reported in many works (Mulligan *et al.,* 2001, Diels *et al.,*2002, Khan *et al,* 2004, to mention some). Most works deal with remediation process which use a) inorganic salts, chelants, pH modifiers, etc., b) use of synthetic surfactant to remove metals form contaminated soils, and c) use of natural surfactants arising from bacteria or plants.

Though points b and c are more related with the objective of this work, it is interesting to mention that soil washing in the case of metal contaminants has been historically developed using inorganic salts, chelants and pH modifiers. For example El-Shafey and Canepa (2003) reported the washing of soils contaminated with Cr (VI) using sulfuric, hydrochloric, acetic and citric acids, EDTA, and NaOH solutions. These authors washed contaminated soils with these solutions and reduced Cr (VI) using a sorbent prepared from rice husk. In another work, Jean et al (2007) employed chelants such as EDTA, citric acid and histidine. They measured the mobilization of Cr from the contaminated soils for particular fractions such as acid extractable, reducible, oxidisable and residual portions of Cr.

Table 2. Some TPH removal efficiencies for three surfactants, when using different doses (as a multiple of the CMC value).

Surfactant	TPH removal (%)			
	0.5 CMC	50 CMC	100 CMC	300 CMC
Soduim lauryl ether sulphate (Texapon 5)	46.4	56.1	63.8	74.2
Sorbitan monoleate polyethoxylated (Canarcel TW80)	41.9	58.4	75.9	83.3
Nonyl phenol polyethoxylated (Surfacpol 906)	52.5	51.9	83.0	88.2

Adapted form Torres et al. (2004)

Moutsatsou *et al.* (2006) reported the use of washing technology for remediation in soils heavily polluted by mining-metallurgical activities. They tested EDTA, hydrochloric, sulfuric and nitric acids to remove Fe, Cu, Zn, Mn, Pb, As and total metals from a contaminated soil. Finally, Neilson *et al.* (2003) reported the use of DTPA, EDTA, NTA, CMCD, oxalic, citric, nitric and acetic acids in the washing of different soils contaminated with lead.

Regarding the works where surfactants are used, it is possible to say that the focus is very similar to those works where the target compounds are petroleum hydrocarbons. Some differences are very clear for the case of soil contaminated with metals. Doong *et al* (1998) studied the removal of Cd-contaminated soils using a nonionic surfactant (Triton X-100), an anionic surfactant (SDS) and a cationic one (CTAB). The authors explored the use of mixtures surfactant + EDTA in the removal of Zn, Pb, and Cd. SDS resulted very successful in the removal of the three metals. Regarding the mixtures, combination of CTAB + EDTA removed the higher amount of Cd, Pb and Zn, though the combination Triton 100+EDTA and SDS + EDTA were very effective in removing the metals. Of course, the initial pH value was determinant in the removal extent. Comparing the Pb removal at three pH values (*i.e.*, 2.3, 7.1 and 12.3), best results were obtained when using an alkaline soil for the case of SDS + EDTA, but for other surfactant + EDTA combinations results were different.

On the other hand, Kornecky *et al.* (1998) used a cationic surfactant for mobilization of Pb from contaminated soils. In general, it is unusual the use of cationic surfactants in soil washing, since these surfactants show a high level of adsorption onto soils. Nevertheless, these authors observed fair Pb removals using 8 cationic surfactants and EDTA. Best results were obtained using lapyrium chloride (LC), dedecyl pyridinium chloride (DPC) and isoteramydopropyl morpholine lactate (ISML). Removals up to 75% of Pb were obtained using ISML. Using only EDTA, removals as high as 95.4% were achieved.

An interesting work is that reported by Nivas *et al.*(1996), where several anionic, nonionic and zwitterionic surfactants were employed in the remediation of subsurface chromium contamination. The zwitterionic surfactant Deriphat-160 resulted very interesting when removing Cr from soil, giving up to 2.8 times the removal achieved when using only water. Other interesting surfactant resulted Dowfax-8390 (an anionic one).

The simultaneous removal of metals and oil fractions is also possible, as stated by Zhang et al. (2007), who reported the removal of Pb and diesel from a contaminated soil using EDTA- and SDS-enhanced washing. These authors found that EDTA washing, followed by SDS achieved the highest Pb removal efficiency, but the sequence SDS washing followed by EDTA is considered as the optimal washing sequence for diesel removal.

Many works deal with the use of natural surfactants for metal removal from contaminated soils. Hong et al. (1998 and 2002), Miller (1995), Juwarkar et al. (2007), and Torres *et al.* (2007a), among other researchers, have explored this research line. Surfactants such as aescein (from horse chestnut tree), rhamnolipids arising from *Pseudomonas* genera and other microorganisms, and saponine (a plant-derived surfactant) have been used for remediation of contaminated soils with successful results. In general, it is possible to say that natural surfactants display two main characteristics that make them very attractive for remediation issues. Since natural surfactants are biological molecules, they are easier to be biodegraded in comparison with some synthetic surfactants. Secondly, doses required to get certain removal levels are lower if compared with natural surfactants. The use of natural-synthetic surfactant mixtures is highly attractive as reported by Torres et al (2007a).

Scaling-Up of SESW Aspects

Washing process at real scale is a combination of several units and process, including washing tank, separation of fractions, and water treatment (See figure 3). There are two basic processes for washing contaminated soils. The more traditional one is that in which soils are placed in contact with the washing solution in a tank where an impeller or a set of impellers provides the necessary energy to maintain the solids suspended during the operation and for the soil washing itself. The other approach includes the use of hydrocyclones, where soil can be washed and separated form the washing solution in only one step. Toebermann et al. (1999), Wilichowski and Wetrther (1995), and Kuhlman and Greenfield (1999) consider the use of these hydrocyclones, but they are more focused on the tank-impeller systems.

For these authors, the heart of the process is the tank and the impeller, though other equipment needs to be considered, such as the transportation bands for soils, filter press, centrifuges, pumps, surfactant solution storage tanks, piping, and others. In the Chemical Engineering theory, this procedure is studied as the suspension of solids. Many authors have worked with suspension of solids, although most of them have developed the theory using model systems (i.e., particles made of plastic, glass, ceramic, or other material) in water. The general purpose of these works is to design an appropriate system with the capability of suspending the solids during the operation and providing the system with the necessary energy to accomplish the required function (washing of soils, in this case). The basic questions in this field are the following. For a given reactor size and shape, which is the best impeller or combination of impellers to get the desired mixing performance? Which is the minimum agitation speed to maintain complete suspension of solids? What is the effect of handling 10, 20, 30, o 40% of solids in the reactor? What is the effect of these parameters on power consumption?

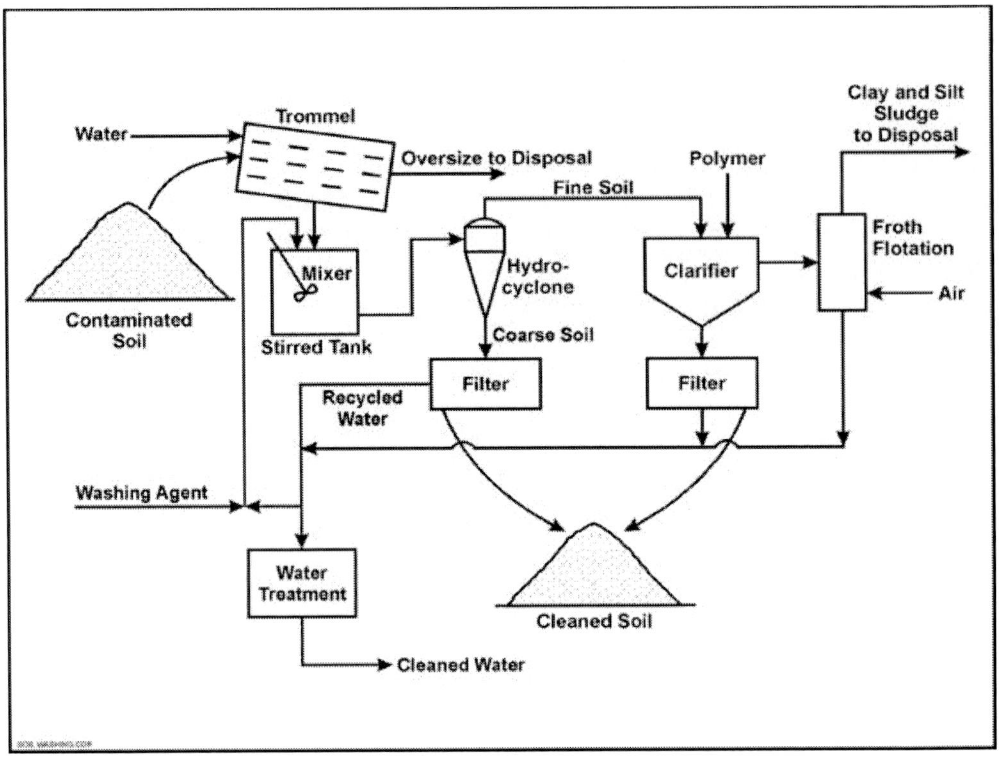

Figure 3. Schematic diagram of a full soil washing process, including washing tank, separation of fractions, and water treatment.

Power consumption is a key factor in the design of the SESW, since it is one of the parameters that can be used for scaling-up a process like this. Uhl and Essen (1986) summarize that equal surface effects, constant blend time, dispersion, suspension of solids, and flow velocity are used as scale-up criteria, and they are all a function of power per unit of volume. Some authors have reported correlations to predict the minimum agitation speed for complete solid suspension (Njs) in agitated vessels (Armentante et al., 1998, among them). Unfortunately, these correlations are limited for two reasons: 1) correlations have been developed for low solid contents (<15%) and 2) correlations have been developed with model solids and fluids.

A better practice is to measure power consumption and Njs in real soil washing operations. It is quite interesting to know the energy consumption developed by the system when using different soil concentrations (up to 40%), different impellers, and different speeds. Our research group has been working in that field, so information useful for scaling-up the SESW process is available (Climent, 2007; Doroteo, 2008; Castillo, 2008).

Figure 4 shows a 3-L system for measuring the agitation speed and power consumption when washing soils contaminated with oil hydrocarbons. The system is capable of working with or without baffles, using different impellers or their combinations (From Castillo, 2008).

Figure 4. Three-liters system for measuring the effect of type of impeller and speed on power consumption when washing soils with surfactant solutions (Photograph: J.A. Castillo).

Regarding to Costs

Remedy screening is the first testing level. It establishes the ability of a technology to treat a specific waste and it generally a low cost study (e.g., $10,000 to $50,000) usually requiring hours to days to be completed (USEPA, 1991). Remedy selection testing is the second level of testing. It identifies the technology's performance for a given site. These studies have a moderate cost (e.g., \$20,000 to \$100,000) and require several weeks to complete. Remedy design testing provides quantitative performance, cost, and design information for remediation and the operational unit. This testing also produces the remaining data required to optimize performance. These studies are of moderate to high cost (e.g., \$100,000 to \$500,000) and require several months to complete. Data obtained from remedy design tests are used to: 1) design the full scale unit, 2) refine cleanup time estimates, and 3) refine cost predictions.

Regarding specific construction and operation costs, Iturbe et al. (2004a) reported costs for an *in situ* flushing process (previously described in the *in situ* flushing section). The volume of soil treated was 41.6 m^3 (38 m^2), equivalent to 69.5 tons of contaminated soil. A rough calculation of the process costs estimated a total of \$104.2/m^3 (\$114/m^2). On the other hand, Iturbe et al. (2004b) reported costs for soil washing of 2,460 tons of contaminated soil by SESW. The cost including labor, building materials, pumping and tubing, ditches' construction, storage tanks, and surfactant costs amounted to a total of \$140/m^3 or \$67/ kg TPH removed.

Soil Washing has some Limitations

Soil washing limitations may be defined as characteristics that hinder cost-effective soil treatment (USEPA, 1991). The main limitations for this technology are:

1. High concentrations of additives may be required in some cases to meet the necessary performance goals.
2. High percentages of silt, clay (more than 30 to 50%) in the soil indicate that soil washing will be unfavorable due to the amount of time and money required to treat this volume of contaminated soil.
3. High humic content in the soil makes separation of contaminants very difficult.
4. Surfactants and other additives are often difficult and expensive to recover from the spent washing fluid for its recycling into the soil washing process.

Conclusions and Perspectives

From the information reviewed in this work it is clear that surfactant enhanced soil washing is a valuable emerging technology to be applied for the restoration of sites contaminated with highly toxic or recalcitrant pollutants such as petroleum hydrocarbons or metals.

As reported, several works dealing with the application of this technology to a wide variety of specific soil conditions and procedures have been reported with the same shared information: the high efficiency and low cost of the procedure.

Despite all this information, more research is required to attain practical knowledge about the use, at real scale, of these methodologies. Experimental evidence about, for example, changes in the hydraulic or mechanical properties of the soil after surfactant-water washing, the use of alternative, novel environmentally friendly surfactants, or the estimation of scaling-up parameters for the soil washing process is still needed.

Wastewater generation during surfactant enhanced soil washing processes is a major problem when the generated wastes contain a complex mixture of the pollutants included in the soil matrix as well as remnants of the surfactant used.

REFERENCES

[1] Ahn C.K., Y.M. Kim, S.H. Woo and J.M. Park (2007) Soil washing using various nonionic surfactants and their recovery by selective adsorption with activated carbon. *Journal of Hazardous Materials*.154(1-3):153-60.

[2] Armentante P.M, Nagamine E.U., and Susanto J. (1998) Determination of correlations to predict the minimum agitation speed for complete solid suspension in agitated vessels. *The canadian cournal of chemical enginieering*. 76: 413-419.

[3] Berselli, S., Milone G, Canepa P, Di Giola D, and Fava F. (2004) Effects of cyclodextrins, humic substances, and rhamnolipids on the washing of a historically contaminated soil and on the aerobic bioremediation of the resulting effluents. *Biotechnology and Bioengineering*. 88(1):111-120.

[4] Castillo J.A. (2008) Effect of the soil/surfactant solution ratio over the power consumption and washing efficiency. UNAM. School of Chemistry. Mexico.

[5] Climent M. (2007) Effect of mixing intensity over the efficiency of oil hydrocarbons conatminated-soil washing. Bachelors' Thesis. UNAM. School of Chemistry. Mexico.

[6] Deshpande S., Shiau B.J., Wade D, Sabatini D.A. and Harwell JH. (1999) Surfactant selection for enhancing ex situ soil washing. *Water Research*. 33(2)351-360.

[7] Deshpande S, Wesson L., Wade D., Sabatini D.A., and Harwell J.H. (2000) Dowfax surfactant components for enhancing contaminant solubilization. *Water research* 34(3):1030-1036.

[8] Diels L, van de Lelie N., Bastiaens L (2002) new developments in treatment of heavy metal contaminated soils. Re/Views in Environmental Science and Bio/Technology. 1: 75-82.

[9] Doroteo E. (2008) Effect of some variables over power consumption and oil-hydrocarbons soil contaminated washing process efficiency. Bachelors' Thesis. UNAM. School of Chemistry. Mexico.

[10] Doong R-an, Wu Y-W., Lei W-gang (1998) Surfactant enhances remediation of cadmium contaminated soils. *Water Sceince and Technology*. 37(8): 65-71.

[11] El-Shafey E.I. and Canepa P. (2003) Remediation of a Cr (VI) contaminated soil: Soil washing followed by Cr (VI) reduction using a sorbent prepared from rice husk. *Journal of Ohysics IV France*. 107: 415-418.

[12] Huibers P.D.T., Lovbanov V.S , Fatritzky A.R., Shah D.O. and Karelson M. (1996) Prediction of critical micelle concentration using a quantitative structure-property relationship approach. 1. Nonionic surfactants. Langmuir 12: 1462-1470.

[13] Hong K-J., Choi Y-K, Tokunaga S. Ishigami Y. and Kaijiuchi T. (1998) Removal pf cadmium and lead using asceina as a biosurfactant. *Journal of surfactants and detergents*. 1(2): 247-250.

[14] Hong K-J., Choi Y-K, Tokunaga S. and Kaijiuchi T. (2002) Evaluation of remediation process with plant-derived biosurfactant for recovery of heavy metals from contaminated soils. *Chemosphere.* 49: 379-387.

[15] Iturbe R.,. Flores C, Chavez C., Ramirez A. and. Torres L.G. (2004a) In situ flushing of contaminated soils from a refinery : organic compounds and metal removals. Remediation. Spring: 141-152.

[16] Iturbe R., Flores C., Chavez C., Bautista G. and Torres L.G.. (2004b) Remediation of contaminated soil using soil washing and biopile methodologies at field level. *Journal of Soils and Sediments.*

[17] Juwarkar A.A. , Nair A., Dubey K.V., Singh S.K., and Devotta S (2007) Biosurfactant technology for remediation of cadmium and lead contaminated soils. *Chemosphere.* 68: 1996-2002.

[18] Jean L. Bordas F., Bollinger J-C (2007) Chromium and nickel mobilization froma a contaminated soil using chelants (2007) *Environmental Pollution*. 147: 729-736.

[19] Khan FI., Huisain T, Hejazi R (2004) An overview and analysis of site remediation technoilogy. *Journal of Environmental Management*. 71: 95-122.

[20] Kornecky T.S., Brown G.O., Allred B, and Basta N. (1998) Cationic surfactant feasibility for use in removal of lead from soil. *Environmnetal Geosciences*. 5(1): 29-38.

[21] Kuhlman MI and Greenfield TM (1999) Simplified soil washing processes for a variety of soils. *Journal of Hazardous Materials.* 66: 31-45.

[22] Kuei-Jyum C. and Yopung Ch-Ch. (2003) Effects of soil fines and surfactant sorption on contaminant reduction of coarse fractions during soil washing. Journal of Environmental Science and health. Part A. Toxic/Hazardous substances & Environmental Engineering. A38(11): 2697-2709.

[23] Lopez J., Iturbe R. ,Torres L.G. (2004) Washing of soil contaminated with PAHs and heavy petroleum fractions using two anionic and one ionic surfactant: effect of salt addition. *Journal of Environmental Science and Health*. A39(9): 2239-2306.

[24] Miller R.M. (1995) Biosurfactant-facilitated remediation of metal- contaminated soils. Environmental Health Perspectives 103, Supplement 1. Published online.

[25] Moutsatsou A., Gregou M., Matsas D. And Protonotarios V. (2006) Washing as a remediation technology applicable in soils heavyly polluted by mining-metallurgical activities. *Chmosphere*: 1632-1640.

[26] Mulligan C.N., Yong R.N,, Gibbs B.F. (2001) Surfactant-enhanced remediation of contaminated soil: A review. *Engineering and Geology* 60(1-4), 371-380.

[27] Neilson J.W., Artiola J.F., Maier R.M. (2003) Characterization of lead removal from contaminated soils by non-toxic soil-washing agents. *Journal of Environmental Quality.* 32: 899-908.

[28] Nivas B.T., Sabatini D.A., Shiau B-J, and Harwell J.H. (1996) Surfactant enhanced remediation of subsurface chromium contamination. *Water Research*. 30(3): 511-520.

[29] SEMARNAT (2000)

[30] Toeberman Ch., Werther, J., Rosenkrauz, J., Gruhn, G. (1999) Flowsheet simulation of soil-washing process. *Chemical Engineering Technology*. 22(7): 558-567.

[31] Torres, L.G., Orantes J.L., Iturbe R. (2004) Critical micelle concentrations for three surfactants and their diesel-removal efficiencies in petroleum-contaminated soils. *Environmental Geosciences*. 10(1): 28-36.

[32] Torres L.G., Aguirre A.L., Verdejo A., Iturbe R. (2005) Enhanced soil-washing treatment for soils which are highly contaminated with crude oil. In: Ecosystems and Sustainable Development V. WIT Transactions on Ecology and the Environment. 81: 541-550.

[33] Torres L.G., Orantes J.L., Iturbe R. (2006) Biodegradation of two non-ionic surfactants used for in situ flushing of oil-contaminated soils. *Tenside Surfactant Detergents* 43 (5), 251–255.

[34] Torrres L.G.,. Zavala P, Beltran M., Vaca M., Iturbe R. (2007a) Combination of natural gums and synthetic surfactants for washing of a soil highly contaminated with crude. *Environmental Geosciences*. 14(1): 1-9.

[35] Torres L.G., Climent M., Saquelares J., Bandala E.R., Urquiza G., Iturbe R. (2007b) Characterization and treatability of a contaminated soil from an oil exploration zone. *International Journal of Environmental Science and Technology. 4*(3), 311-322.

[36] Torres LG, Lemus X., Iturbe R. (2007c) Do the characteristics of crude oil in contaminated soils affect its removal by washing? *Land Contamination and reclamation*. 15(4): 391-399.

[37] Uhl V.W., Von Essen J.A. (1986) Equipment for agitated liquids. In : Mixing. Theory and Practice. Uhl V.W.,Gray J.B., Editors, .Academic Press Inc. USA.

[38] Urum K., Pekdemir T.. (2004) Evaluation of biosurfactants for crude oil contaminated soil washing. *Chemosphere* 57: 1139-1150.

[39] Urum K, Pekdemir T., Ross D. , Grigson S. (2005) Crude oil contaminated soil washing in air sparging assisted stirrer tank reactor using biosurfactants. *Chemosphere* 60:334-343.

[40] USEPA (1991) Guide for conducting treatability studies under CERCLA:Soil washing. Interim guidance. EPA/540/2-9/020A. September 1991.USA.

[41] Vreysen S, Maes A. (2005) Remediation of a diesel contaminated, sandy-loam soil using low concentrated surfactant solutions. *Journal of Soils and Sediments*.5(4): 240-244.

[42] J.-Chr. Toebermann, Werther J., Rosenkranz J., Gruhn G. (1995) Flowsheet Simulation of Soil-Washing Processes. *Chemical Engineering & Technology*. 22 (7): 558 - 567

[43] Zhang W, Tsang D.C.W. and Lo I.M.C. (2007) Removal of Pb and MDF from contaminated soils by EDTA- and SDS-enhanced washing. *Chemosphere*.66 2025-2034.

[44] Zhou Q, Sun F., Liu R. (2005) Joint chemical flushing of soils contaminated with petroleum hydrocarbons. *Environment International*. 31(6): 835-839.

Author List

Luis G. Torres. Unidad Profesional Interdisciplinaria en Biotecnología. Instituto Politécnico Nacional. Av. Acueducto s.n. Colonia Barrio la Laguna Ticomán. México 07340 D.F. México. Tel +52 55 57296000 ext 56367 and 56342 LTorresB@iingen.unam.mx (Editor).

Erick R. Bandala. Departamento de Ingeniería Civil y Ambiental. Universidad de Las Américas-Puebla. Sta. Catarina Mártir, Cholula 82720 Puebla, México. Tel: +52 222 2292652; Fax: +52 222 2292000 ext. 4199. erick.bandala@udlap.mx. (Editor).

Alejandrina Castro. Instituto de Ingeniería, Universidad Nacional Autónoma de México, Environmental Engineering Coordination, Cleaning of Soils and Aquifers Study Group, Apartado Postal 70-472 Coyoacan 04510 México City, México.

Diana Santa. Instituto de Ingeniería, Universidad Nacional Autónoma de México, Environmental Engineering Coordination, Cleaning of Soils and Aquifers Study Group, Apartado Postal 70-472 Coyoacan 04510 Mexico City, México.

Rosario Iturbe. Instituto de Ingeniería, Universidad Nacional Autónoma de México, Environmental Engineering Coordination, Cleaning of Soils and Aquifers Study Group, Apartado Postal 70-472 Coyoacan 04510 Mexico City, Mexico

Dr. Wini Schmidt. German Agency for Technical Cooperation (GTZ). Av. Revolución No. 1425, nivel 30. Col. Tlacopac, San Ángel. México, D.F. México.

Stephen Leharne. School of Science, University of Greenwich, Pembroke, Chatham Maritime, Kent, ME4 4TB, UK.

José E. Santos Jallath. Departamento de Explotación de Minas y Metalurgia. Facultad de Ingeniería, UNAM

José A. Ramos-Leal. Instituto Potosino de Investigación Científica y Tecnológica, A.C. (IPICYT), C. Presa San José # 2055, Lomas 4ª. Sección, CP 78216, San Luis Potosí, SLP, México.

J.J. Carrera-Hernández, Instituto Potosino de Investigación Científica y Tecnológica, A.C. (IPICYT), C. Presa San José # 2055, Lomas 4ª. Sección, CP 78216, San Luís Potosí, SLP, México.

Rosa Ma. Flores. Instituto de Ingeniería, Universidad Nacional Autónoma de México, Environmental Engineering Coordination, Cleaning of Soils and Aquifers Study Group, Apartado Postal 70-472 Coyoacan 04510 México City, México.

José A. Hernández-Espriu. División de Ingeniería en Ciencias de la Tierra, Facultad de Ingeniería, UNAM. Facultad de Ciencias Geológicas, Universidad Complutense de Madrid.

José A. Reyna-Gutiérrez. División de Ingeniería en Ciencias de la Tierra, Facultad de Ingeniería, UNAM

INDEX

B

C

D

H

I

J

K

L

N

O

P

Q

R

S

T

U

V